MW01224099

Biotechnology, Agriculture and the Developing World

Biotechnology, Agriculture and the Developing World:

The Distributional Implications of Technological Change

Edited by

Timothy Swanson
Chair of Law and Economics, University College London, UK

Edward Elgar
Cheltenham, UK • Northampton, MA, USA

Published by
Edward Elgar Publishing Limited
Glensanda House
Montpellier Parade
Cheltenham
Glos GL50 1UA
UK

Edward Elgar Publishing, Inc.
136 West Street
Suite 202
Northampton
Massachusetts 01060
USA

A catalogue record for this book
is available from the British Library

Library of Congress Cataloguing in Publication Data

Biotechnology, agriculture and the developing world : the distributional implications of technological change / edited by Timothy Swanson.
p. cm.
Includes bibliographical references (p.).
1. Agricultural biotechnology—Developing countries. 2. Agriculture—Technology transfer—Developing countries. 3. Agrobiodiversity—Developing countries. I. Swanson, Timothy M.
S494.5.B563B536 2002
631.5′233′091724—dc21 2001033999

ISBN 1 84064 679 9

Printed and bound in Great Britain by MPG Books Ltd, Bodmin, Cornwall

Contents

List of figures vii
List of tables viii
Acknowledgments x
List of contributors xi

Introduction

1. Biotechnologies and developing countries: how will the anticipated industrial changes in agriculture affect developing countries? 3
Timothy Swanson

Part I Setting the Scene: the Framework for Considering Biotechnology's Impacts

2. Population growth and agricultural intensification in developing countries 25
Nadia Cuffaro
3. The impacts of GURTs: agricultural R&D and appropriation mechanisms 44
Timothy Swanson and Timo Goeschl
4. Agricultural biotechnology and developing countries: proprietary knowledge and diffusion of benefits 67
Charles Spillane

Part II A Case Study on Terminators: the Impacts of Biotechnologies on Benefit Distribution

5. The impact of terminator gene technologies on developing countries: a legal analysis 137
William W. Fisher
6. Impact of terminator technologies in developing countries: a framework for economic analysis 150
C.S. Srinivasan and Colin Thirtle

7. The impact of GURTs on developing countries: a preliminary assessment 177
Timothy Swanson and Timo Goeschl

8. Forecasting the impact of genetic use restriction technologies: a case study on the impact of hybrid crop varieties 198
Timo Goeschl and Timothy Swanson

Part III Biotechnology and Biodiversity: the Impacts of Biotechnologies on Conservation of Genetic Resources

9. Key issues in using molecular techniques to improve conservation and use of plant genetic resources 219
Carmen de Vincente, Toby Hodgkin and Geoffrey Hawtin

10. Biotechnology and traditional breeding in sub-Saharan Africa 230
Vittorio Santaniello

Conclusion

11. Policy options for the biotechnology revolution: what can be done to address the distributional implications of biotechnologies? 249
Timothy Swanson and Timo Goeschl

Index 271

Figures

2.1 Global population growth 25
2.2 The induced innovation model 27
2.3 *Bt* varieties 36
2.4 *RR* varieties 36
3.1 Vertical industry for plant breeding 47
5.1 Economic effects of profit-maximizing behaviour by the patentee of a new plant variety 140
5.2 Economic impact of partial price discrimination 144
6.1 Appropriability and private research investment in plant breeding 162
6A.1 Undiscounted quasi-rents for different seed replacement practices 176
6A.2 Discounted value of quasi-rents under different seed replacement patterns 176
8.1 Loss multiplier as a function of the diffusion rate 204
8.2 Yield histogram and yield statistics for the year 2020 for 86 developing countries in the baseline scenario 207
8.3 Yield histogram and yield statistics for the year 2020 for 86 developing countries in the GURTs scenario 207
8.4 Comparison of yields under the use restriction and baseline scenarios, developed countries, 2000–2020 209
8.5 Comparison of yields under the use restriction and baseline scenarios, China, 2000–2020 210
8.6 Comparison of yields under the use restriction and baseline scenarios, Ethiopia, 2000–2020 210
8.7 Comparison of yields under the use restriction and baseline scenarios, Tanzania, 2000–2020 211
11.1 Cycles of production and exchange without and with GURTs 263

Tables

2.1 Projected growth rates of population, aggregate demand and production 34
3.1 Private research and development activity (wheat case study) 54
3.2 PBR titles granted in three developing countries, 1968–94 55
3.3 Average R&D expenditure of plant breeding companies in Argentina 55
3.4 Real agricultural R&D spending in OECD countries, 1971–93 57
3.5 Real agricultural R&D spending in developed and developing countries, 1971–91 58
6.1 Summary of innovator's discounted quasi-rents for different scenarios 160
6.2 Seed sales, private plant breeding and trends in seed prices and yields of major field crops, USA (1975–92) 161
7.1 Classification of impact groups with respect to GURTs 182
7.2 Criteria for classification of impact groups 183
7.3 Developing countries grouped according to GURT impacts 186
7.4 Real agricultural R&D spending in developed and developing countries, 1971–91 191
7.5 Rate of R&D diffusion to countries dependent on funding (under different technology management systems and levels of public R&D spending) 191
7.6 Baseline and other scenarios for appropriation systems 192
8.1 Regressions for diffusion of innovations in different crops 203
8.2 Country-specific simulation parameters 208
10A.1 Africa: activities in biotechnology, by country 245
11.1 R&D structure and land use patterns for existing agricultural R&D systems 253
11.2 Average difference between crop yields of developing countries and developed countries, 1999 257
11.3 Real agricultural R&D spending in developed and developing countries, 1971–91 258
11.4 Impact of diffusion rates on the rate of productivity growth for resource-poor farmers 260
11.5 Imputed within-zone yield gaps for Latin America, 1996 261

11.6 Profile of advanced and resource-poor sectors in a developing country with respect to key requirements in the cycle of agricultural production and exchange 262
11.7 Rate of R&D diffusion to countries dependent on funding (under different technology management systems and levels of public R&D spending) 264
11.8 Baseline and other scenarios for appropriation systems in agriculture 265

Acknowledgments

Much of this work had its origins within a research grant from the UK Department for International Development on the distributional implications of genetic use restriction technologies. We would like to acknowledge this contribution, and the guiding hand of Robert Carlisle of that department. A workshop was hosted at the International Plant Genetic Resources Institute, Rome, Italy, in May 2000. We would like to thank Geoffrey Hawtin and Pablo Eyzaguirre for their hospitality, and for managing an exceptional event. Funding for the workshop was received from both the UK Department for International Development and the European Science Foundation. We gratefully acknowledge the financial support from all of these agencies, which made the production and distribution of this volume possible. I would also like to thank all of those at Edward Elgar who contributed to the production of the volume. They do excellent work.

Contributors

Nadia Cuffaro, University of Cassino, Italy

William W. Fisher, Professor of Law, Harvard University, USA

Timo Goeschl, Department of Land Economy, Cambridge University

Geoffrey Hawtin, The International Plant Genetic Resources Institute, Rome, Italy

Toby Hodgkin, The International Plant Genetic Resources Institute, Rome, Italy

Vittorio Santaniello, Tor Vergata University, Rome, Italy

Charles Spillane, Institute for Plant Biology, University of Zurich, Switzerland

C.S. Srinivasan, Department of Agricultural and Food Economics, University of Reading, UK

Timothy M. Swanson, Chair of Law and Economics, University College London

Colin Thirtle, Professor, Imperial College, London

Carmen de Vicente, The International Plant Genetic Resources Institute, Rome, Italy

Introduction

1. Biotechnologies and developing countries: how will the anticipated industrial changes in agriculture affect developing countries?

Timothy Swanson

BIOTECHNOLOGY AS AN INDUSTRIAL MOVEMENT

Biotechnological advances in agriculture refer to two distinct forms of change: technological and industrial. The technological changes involved refer to advances in genetic marking and transfer methodologies, which enable new and dramatic changes in life forms in short amounts of time. These technological changes and their management implications are the subject of a separate volume (Swanson, 2001). The changes we investigate in this volume concern the dramatic changes to be expected within the industry of modern agriculture, rather than its biology.

These changes in industrial structure are to be anticipated because they will address a difficult problem that has plagued the agricultural industry for decades: the problem of the 'durable goods monopolist'. The durable goods monopolist is faced with the paradox that the sale of its unique product introduces competitors into its own market on account of the good's 'durability'. This refers to the ability of purchasers to store and then to resell the good. Hence the owner of a durable innovation sits unhappily on the horns of a dilemma. If the innovator commences marketing the new product, the purchasers of the product are able to enter into competition with it. If it does not sell the product, however, it is unable to reap any returns from its investments in innovation. In either event, the producer sees little possibility of generating returns from investments in innovative goods that are able to be resold.

Agricultural innovations fit all too well into this paradigm. Innovative plant varieties are not only able to be resold, they may be reproduced identically and resold in large quantities. And this may be done with little investment other than the land required to plant the purchased seed. Therefore the sale of an innovative plant variety introduces the prospect of a competing producer, not just a competing seller. Since many plant varieties are geared towards asexual reproduction, the quality of the competing products is assured through cloning. The innovator must consider the possibility that any sale of the innovative plant variety contains an implicit licence to enter into its replication and production.

This implies that the producer must either price its innovative product to include this implicit licence or attempt to exclude purchasers from reselling the product if they do not have such a licence. Either option is costly. The sale of innovative plant varieties bundled with a licence to reproduce/resell them would imply a very high price, and hence exclude those users who wish only to make a single season's purchase decision. This would require the innovator to recoup its investment in a single season's sales, to a limited number of purchasers. On the other hand, if the plant variety is sold without the licence to reproduce, this implies the enforcement of rights that are held privately. This involves very costly monitoring and enforcement activities, mostly on the part of the state concerned.

Imagine the situation of the plant breeder who is selling new plant varieties that are unprotected. Suppose that the plant breeder recognizes that a widely-used variety of wheat is facing increasing problems of pest infestation. It then isolates the characteristic of a particular wheat variety that is resistant to the current pest problems and develops the techniques that incorporate this characteristic into the widely-used variety. After trials, it grows the new plant variety on sufficient land area to allow it to market the seed for the crop. Finally, it is able to offer the new, resistant variety of wheat to farmers. It is offered on the market at a substantial premium over the wheat variety that is in decline, but with the promise of higher yields. Suppose that a minority of farmers purchase the new variety the first year, and do indeed find that it increases their yields substantially over the non-resistant variety. Then, at the end of the year, these farmers are able to withhold some of their crop for replanting the following year, and even for sale to a few of the neighbouring farmers. How much will farmers charge for the seed? If there is a sufficient number of them, they will charge only the opportunity cost of the seed to them, that is, a little more than the cost of a bushel of wheat on the market. So, after only one year of a partial return on its investment, the plant breeder finds itself in competition with a multitude of sellers of its own product, all willing to sell at a competitive price. In this way the private market provides little prospect of a return to investments in the necessary research and development on new plant varieties.

Hence the plant breeding industry in agriculture is faced with a very serious variant of the durable goods monopoly problem. But why should we – the consumers of the world – care if this problem is resolved? That is, is it not beneficial to consumers from all parts of the world for monopolies to be broken down as rapidly as possible? The problem here lies in the need to provide returns to the investments by the producers of new plant varieties, so that they will continue to work on solutions to recurring problems of pest resistance and yield decline. The widespread use of a small number of plant varieties – as often occurs in modern agriculture – implies that pest infestation and epidemics will increase over the time of use. On account of this, a widely-used plant variety tends to have a commercial life of only five to seven years (Goeschl and Swanson, 2001). This means that plant breeders must be working continuously on producing new, innovative varieties in order to replace the old. If these innovations are not coming online when the old are in decline, modern agriculture will become impracticable.

The plant breeding sector must make substantial investments in research and development before a new plant variety can be developed and released. The only way in which it can acquire compensation for these investments in the private marketplace is by means of some sort of exclusive marketing rights in the resulting innovative varieties. Simply selling the seed in competition with other sellers would generate no return to the innovation; it would generate a return only to the land planted. The rationale for intellectual property rights is that they provide a limited monopoly right in order to generate returns to investments in the production of innovations.

It is precisely this scenario that generates the plant breeder's dilemma regarding investments in the plant breeding industry. Without a clear means of appropriating the value of its investments, the firm has little incentive to pursue research and development in this field. Without this research and development, modern agriculture has little capacity for continuing into the indefinite future. The industry has been lobbying for an enforceable method of appropriating the value of innovation over the past few decades. So-called 'Plant Breeders' Rights' came into being originally in Europe and in the USA more than a half-century ago (see Fisher, this volume). These rights were made more uniform through the formation of the Union pour la Protection des Obtentions Végétales (UPOV) in 1961. These laws vested plant breeders with exclusive marketing rights in registered plant varieties. Analogously to patent systems, the plant breeder was required to register the plant lines used to generate the innovative variety, and to demonstrate that the desirable characteristics of the variety were stable and replicable. Once these requirements were met, the plant breeder was recognized to hold the exclusive right to market that plant variety, and this right was recognized reciprocally across all of the nations within the Union.

The problem with such rights is that, as with other private forms of use of copyright materials, the division of rights between producers and consumers is a fundamental policy decision. The line between permissible 'private use' and illegitimate copyright infringement is difficult to draw, and to enforce with regard to all products that contain reproductive technology. The recent court decision in *Napster* illustrates the difficulties involved in drawing lines between permissible and impermissible reproduction of copyrighted material. In earlier cases (such as *Sony v. Amstrad*) the courts had allowed the sale of reproduction technologies (such as video-cassette recorders) under the rationale that the technologies were intended to foster the legitimate private copying of copyrighted materials. In *Napster*, the court held that the producer (a software company) was aiding the illegal private copying of copyrighted materials. Thus the lines between legitimate and illegitimate private uses of copyright materials are never clear-cut, and always involve a policy determination concerning the division of uses between consumers and producers.

In 1978, the UPOV convention recognized this problem in conferring a limited 'Farmers Right' to the private reuse (replanting) of registered seeds. This right of use was analogous to the 'own use' provision in copyright law. The 'own use' provision allows the purchaser of copyrighted material to make a very limited number of copies of the materials for its own use. This provision has come under fire from the biotechnology industry in recent decades (and was removed from UPOV in 1991) as a means of disguising improper use. At the same time the UN's Food and Agricultural Organisation (FAO) has adopted the International Undertaking on Plant Genetic Resources (IUPGR), which has provided for the reciprocal recognition of both 'Plant Breeders' and 'Farmers' rights. The international undertaking (IUPGR) remains to be implemented, but its adoption clearly indicates that the line between legitimate and illegitimate uses of purchased plant varieties remains under discussion within international fora.

Even more problematic for the plant breeding industry has been the unwillingness of many states to make any effort in the enforcement of plant breeders' rights, even when they exist on the books. This failure was at the heart of the long-running Uruguay Round of GATT discussions, culminating in 1994 with the TRIPS (Trade Related Intellectual Property Rights) Agreement. Article 27 of TRIPS provides that all member states of the World Trade Organisation (WTO) must 'provide for the protection of plant varieties'. WTO membership now confers upon a state the responsibility to enact plant variety protection legislation (such as that consistent with UPOV), as well as the responsibility to enforce that legislation within its jurisdiction.

Why is it that different states feel differently about plant breeders' rights? The reason for many countries having little incentive to enforce these rights is the asymmetry between the nationality of most rights holders and the nationality of most users. The plant breeding industry is concentrated in a few firms, and these firms are large multinationals based primarily in Europe and the USA. This has been true for many years now (Juma, 1988) and, for this reason, there are only a very few nations that view the enforcement of these 'producers' rights' against the 'users' as a governmental function that generates direct benefits for their own citizens. This asymmetry between the nationalities of producers and users renders distinctly perceived national interests regarding the question of the division of rights between producers and users. Since the line between producers and users rights is never drawn perfectly clearly, this ambiguity allows the individual state some flexibility in its own interpretation and enforcement of the law. Different countries have enforced the laws differently, depending on their interpretation of their own state interests.

Even with the best of intentions, the allocation of scarce resources to the monitoring and enforcement of the legal rights of foreigners is a difficult task to give high priority. In the poorer countries where governmental resources are under very heavy pressure, it is readily apparent why this object is given little or no weight whatsoever. Hence, in many if not most countries without a plant breeding industry, the enforcement of plant breeders' rights has been problematic to say the least.

To this point we have been able to identify the fundamental problem of appropriation afflicting the plant breeding industry, and why this problem was generated in part by reason of the perceived self-interest of those countries without a plant breeding industry. The plant breeding industry had made long-standing and expensive efforts at acquiring enforcement efforts, but the fundamental asymmetry within the industry kept these efforts from coming to fruition.

This points to the need for technological, rather than legal, efforts at enforcement of plant breeders' rights. The plant breeding industry was able to achieve a technological solution to the enforcement problem with respect to a few of its products. For a few plant varieties (including maize and sorghum) sexual reproduction is the norm, and this affords an in-built enforcement strategy. When the second generation of two distinct lines (first generations) is produced through hybridization, this second generation will exhibit traits that neither the first nor any subsequent generation will display. This is because the second/hybrid generation will again reproduce sexually, mixing the gene pool and creating a distinct set of characteristics. Thus innovative hybrid crop varieties were marketable by their breeders without the threat of resale by the consuming public. In economic terms, the hybrid varieties were not 'durable goods'. The fundamental problem of plant breeders was solved, more or less, for these particular crop varieties, by reason of this in-built difference in biology. Hybridization was the first technology used as a method of 'use restriction'.

It would be a very short-sighted industry indeed that did not appreciate the commercial importance of this difference, and did not consider the possibility of extending this characteristic to other crop varieties. The advent of new biotechnologies reduced the barriers between various species (including plant varieties). The possibilities for transferring desirable traits between species were expanded as never before. An obvious 'next step' within the plant breeding industry was to investigate the translocation of these use restriction technologies through genetic transference.

The area of biotechnological research and development focused on the problem of appropriability has been referred to as the field of 'genetic use restriction technologies' or GURTs. GURTs come in two distinct forms, at least theoretically. Variety-based GURTs (V-GURTs or 'Terminators') are plant varieties that are not reproducible in any way by the purchaser. The basic idea is to create a seed that will generate the desired plant variety that itself generates sterile seed. Thus, with V-GURTs, the purchaser acquires the innovative plant variety without acquiring the technology for reproducing any part of the plant. Trait-based GURTs (or T-GURTs) are plant varieties with the potential for innovative traits, but requiring the application of a complementary product (an initiator) that causes the trait to come to fruition. With T-GURTs, the purchaser acquires the reproductive technology for the standard plant variety, but must purchase the complementary product to acquire the benefits of the innovation. It should be noted that neither technology is yet in commercial use, but both are feasible.

GURTs is the area of research concerned with the technological resolution of the problem of appropriability that so severely afflicts the plant breeding industry. It is the logical intersection of three distinct and dynamic processes: agricultural, legal and technological. Agriculture has evolved into an enterprise heavily dependent on research and development, represented by the cycling of widely-planted varieties subjected to increasing pest and pathogen problems. The international legal system has struggled to create an incentive system capable of rewarding such research and development investments, on account of the asymmetry between the nationalities of investors and users. Technology has stepped up to fill this gap by evolving the means through which use restrictions might be built into most crop varieties.

Thus a very important part of the biotechnology revolution in agriculture concerns this fundamental change in the industrial structure of agriculture. Biotechnologies are pursued for profits by the private sector, and the solution of the appropriability problem in plant breeding is an important potential source of increased profitability. Much early effort has gone into developing the technologies for enhancing the appropriability of returns from innovation, rather than into innovation itself. Thus the pursuit of GURTs is a fundamentally important part of the biotechnology industry.

Is this a socially desirable objective? Will GURTs advance the general cause of humanity? Or are these technologies biased towards some segments of global society at the expense of others? These are the issues at the core of this volume. It is an enquiry into the distributional impacts of a technological change of this nature. In particular, it is an enquiry into the anticipated impacts of such a technological change on the poorest segments of the world's population. How will they benefit from advances in appropriability that primarily accrue to the high technology sectors of the most developed countries in the world?

It is important to recognize that *in the abstract* and *all other things being equal*, the enhanced appropriability of the benefits from innovation can only represent a social advance. Enhanced appropriability means only that the creator of an innovation is able to acquire a greater share of the benefits that it has generated for society. The innovator is not able to capture one iota more than the consumer is willing to pay for the innovation. To the extent that the innovator perceives that it will be able to capture more benefits, the innovator will invest in generating more innovations. In general, there is a widely-held belief amongst economists that the benefits of innovation are never fully appropriated, and hence increased rates of innovation will usually generate increased social benefits.

The problem with this argument lies in the fact of inequality. The initial conditions under which these technological innovations are occurring are far from equal. As mentioned previously, the distribution of the capacity for innovation is highly skewed. Thus the distribution of initial benefits from enhanced appropriability is similarly skewed. The distribution of extremely poor peoples is also highly skewed, and the potential for disastrous results is therefore also very skewed. The distribution of governmental resources is highly skewed, and hence the capacity to deal with new problems is highly skewed. This diversity of initial conditions means that there can be a highly divergent set of impacts on different parts of humanity. One object of this volume is to sort out the various initial conditions that matter, and to assess how these conditions will cause impacts to differ.

In addition, these industrial changes are not occurring within a vacuum but against a long history of negotiations, as set out above. The existing distribution of benefits from global agriculture – between industry, farmer and consumer – sits within a delicate equilibrium. The existing legal structure, the national enforcement strategies and the international negotiations concerning these matters all represent the various forces shaping the distribution of these benefits. The injection of an industrial strategy that reshapes this distribution will have significant social impacts: locally, nationally and internationally.

Therefore this volume attempts to set out the manner in which these industrial changes will affect the world by reference to the actual conditions under which they are occurring. The various authors have presented varying perspectives on how biotechnological changes will have an impact on the distribution of benefits and rights, relative to the current regime. They do not speak with a single voice – a wide range of perspectives on the problem is evident – but they all indicate the fundamental nature of the changes taking place within the agricultural industry. The original papers resulted from part of a workshop held at the International Plant Genetic Resources Institute in May 2000, sponsored by the UK Department for International Development.

In the remainder of this introduction the three parts of the volume are outlined. The individual papers are reprised briefly for those who are most time-constrained, but for those able to pursue the topic in some detail it is recommended that the reader turn to the contributions themselves.

SURVEYING THE IMPACTS OF BIOTECHNOLOGY: TECHNOLOGY, AGRICULTURE AND DISTRIBUTION

Part I of the volume presents both the abstract economic framework and the more specific institutional framework for the ensuing discussion. It sets forth the basic framework within which this industrial change is occurring.

The chapter by Nadia Cuffaro surveys the literature on the relationship between institutional status and technological change in agricultural development. Her survey commences with the Boserup model of induced technological change, in which technology responds continuously to real underlying factors such as population growth. Then the limits of this model are indicated by the Hayami and Ruttan (1985) framework for the discussion of the role of institutional adjustment in induced technological change. This model states that the impact of institutional differences is to cause substantial and differential time lags in the responsiveness of distinct societies to changed conditions. Thus the ability of individual countries to respond to changes in fundamental conditions depends on their institutional make-up. For our purposes, this indicates that the differential ability to absorb technological change will also depend on differential national investment positions. For this reason it is to be anticipated that different countries will benefit differentially from the same technological change.

The importance of this chapter is that it establishes the basic economic framework within which this problem may be considered. It makes clear that agriculture is an industry that responds to societal demands, like any other, and so technological changes are induced by societal needs. At the same time, the capacities of individual societies to benefit from induced technological change will depend on the endowments and investments of those societies. These differences may be cultural, physical or institutional, but the more different countries may be, the more uneven will be the impacts of the technological change. Some societies will be far more able to take on board the changes than others, and the Ruttan–Hayami model indicates that for this reason the diffusion of technological change will occur at differing rates.

Cuffaro makes clear that there is an unevenness to be anticipated from the impact of the same technological change flowing across countries with different characteristics (cultural, physical, institutional). She demonstrates that technological change matters, when all things are not equal. The following chapter, by Swanson and Goeschl, establishes the institutional background against which technological change in agriculture is occurring. It demonstrates that agricultural research and development is a system that makes use of factors – from many if not most parts of the world – in producing its outputs, and that it is wrong to think of the plant breeding sector as a 'stand alone' entity. These outputs are then widely used in modern agriculture, as part of a comprehensive system of agriculture, and the benefits are distributed in accordance with various arrangements and negotiations. A change in industrial structure must be assessed against the background of this institutional structure.

It is important to emphasize that the R&D industry in plant breeding, although centred in the developed world, relies heavily on inputs from the developing. The production of a new plant variety involves, at a minimum, human inputs (scientists), capital inputs (land, laboratories) and natural inputs (diverse genetic resources). While the former are primarily generated in the developed world, the latter are often sourced in the developing. Production function studies have estimated that diverse genetic resources provide nearly a third of the contribution required for the production of new plant varieties (Evenson and Gollin, 1995). Therefore it is probably incorrect to claim that the final segment of the R&D industry – centred in the developed world – is capable of generating these innovative varieties by means of its efforts alone.

The present system for distributing the benefits of agricultural R&D implicitly recognizes the role of the various parts of the vertical industry in generating a plant variety. The concept of 'plant breeders' rights' has long been attenuated by the reciprocal recognition of 'farmers' rights'.

Irrespective of the individual levels of these contributions, the important characteristic of this arrangement is that it recognizes that plant breeders' rights should be attenuated. The plant breeder, as the entity at the end of the industry pipeline, is the only part of the industry able to claim an exclusive marketing right *vis-à-vis* consumers. However, the recognition of the need for this exclusive right need not necessarily imply that the plant breeder is the only part of the industry entitled to a return from R&D (Swanson and Goeschl, 1999). In short, the enhancement of appropriation should not be used to justify the absence of rent sharing within the R&D industry.

To some extent the entire matter is confused because some of the consumers are also suppliers within the industry. Farmers both supply the R&D process (with genetic resources) and then also purchase its ultimate outputs. It is important that the benefits from the R&D process be appropriable (at the end of the pipeline, that is, by the plant breeder) but it is equally important that the contributions of the various factors of production be compensated. In the past these two separate problems were confounded, and farmers came to expect to receive their share of modern agriculture's benefits within the production process. Biotechnology represents a revolutionary change, both in the enhancement of appropriation and potentially in the distribution of rents after appropriation, and it is against this background that it must be judged.

An obvious factor in all this is the role of the public sector in effecting both technological change and the distribution of its benefits. In the past the public sector has tolerated reduced appropriability in part because it enabled a wider distribution of the benefits from modern agriculture. Legal movements have been slowly making inroads against the status quo, in return for the recognition of some sort of 'farmers' rights' and also for public investment in the transfer of technology. This public investment in technology transfer has taken the form of international investment in the Consultative Group on International Agriculture Research and its various research stations around the world (CIMMYT for maize research, IRRI for rice research). It has also taken the form of various bilateral and multilateral investments in 'national agricultural research' stations across the world. One of the most important functions of such investments has been to enhance the rate of diffusion of new technologies from the frontier states (in the developed world) to the needs and uses of the developing countries.

This brings us back to the point of the Cuffaro chapter. Technological change plays an important role in distributional considerations, only to the extent that institutional or other differences influence the rate at which new technologies diffuse across countries. If all countries had the same capacities to absorb new technologies, diffusion of benefits would occur more or less simultaneously. There would be no distributional implications to technological change.

The public sector investments in technology transfer could be used as an alternative mechanism for counteracting the distributional implications of biotechnologies. Even if innovators were able to capture a greater share of their innovation's benefits, the public sector investment would then provide other countries with the capacity to observe and to understand the information embodied within the innovation. Then the other countries would be able to reproduce that information in an innovative form that most suited the situation of that country. Hence public investments in 'technology transfer' can act as a means for encouraging the rate of 'trickle down' from the technological frontier. Technologies matter less, in distributional terms, if their rate of diffusion is rapid.

In Chapter 4, Charles Spillane surveys the current state of play in regard to the public and private roles in agricultural biotechnology. He demonstrates the many different fora within which the legal and institutional issues have been negotiated, and how these various agreements affect both biotechnology and the distribution of its benefits. The chapter does an excellent job of summarizing the various discussions that have occurred about the respective roles of the private and the public sector in agricultural R&D, and it sets forth a wide range of policy alternatives that have been discussed in this regard.

The three chapters together set out the economic, institutional and policy backgrounds against which this biotechnological change is occurring. They make clear that there are fundamental forces involved, but that institutions and policies determine the outcome.

THE IMPACTS OF GURTs: A CASE STUDY ON THE DISTRIBUTIONAL IMPLICATIONS OF TECHNICAL CHANGE

Part II of the volume is a case study on the predicted impacts of genetic use restriction technologies on the structure of global agriculture. This is the core of the volume. As mentioned before, GURTs (or, as they have been dubbed by the popular press, 'Terminators') are an entirely predictable consequence of the biotechnology revolution. Firms pursuing profits are going to be as interested in enhancing the appropriation of the benefits they generate as they are in expanding the range and magnitude of those benefits. Private firms have created GURTs as a mechanism for translating the already successful methods of appropriation used in hybrid varieties to asexually reproducing crop varieties. This change is problematic to the extent that GURTs are more a distributional and less of an efficiency-enhancing phenomenon. The efficiency and distributional differences introduced by the technology are outlined in the four chapters in Part II.

Chapter 5, by William Fisher, indicates the basic nature of the GURTs technology. It enables the producer to sell its innovative product without selling the reproductive technology bundled along with it. This allows the price of the seed to be set at the 'single use price', while allowing the purchasers individually to elect the number of years in which to purchase the product. This means that some purchasers may choose to purchase the innovation for use in a single year, while others may choose to use it each year for a number of years. Others may disdain the innovative feature, and elect not to purchase the variety at all. In the abstract, such a change in marketing technology can only be to the benefit of both producers and consumers. This is because it enables the finer segregation of the market, and allows for the specific targeting of individual user's needs. Users may decide on an annual basis whether they are willing to pay the price for the innovative feature (Fisher, this volume).

The problem with this analysis is that it avoids the issue of the available alternatives. In the first years of use of GURTs, the consumer has a clearly welfare-enhancing choice. It makes use of the freely available standard plant variety, or it makes use of the standard plant variety with the innovative trait imbedded within it. The user makes the decision whether, given individual conditions, it is willing to pay the market price for the innovative trait or not. If the user is willing to purchase the use of the trait for that year, then it clearly must be welfare-enhancing for it to do so. In this regard, GURTs may be analogized to the sale of an annual licence for the use of new software (the innovative trait), and the consumer is allowed the individual choice on whether to acquire the licence or not.

In the case of plant varieties, the software and the hardware become commingled over time. If the plant breeding industry introduces traits only within the context of GURT varieties, then over time the freely available standard variety may come to be something very unlike the variety into which the innovative traits are imbedded. That is, the proprietary traits may be allowed to accumulate within the commercial sector, without allowing their diffusion into the public arena. Then the commercial breeders will be able to work with the commercial 'hardware' (by paying for licences for one another's innovations) while the public sector breeders (individual farmers, universities, government researchers) may be left with antiquated varieties as their alternatives. Within five or ten years, there might be no real alternative to the use of the GURT varieties, because the hardware within the public sector would be without a decade's worth of developments. Then users would become wholly dependent on the plant breeding sector for their seed, unlike the situation at present, where 80 per cent of farmers in developing countries use retained seed (Fisher, this volume).

GURTs provide even more substantial protection than would perfectly enforced intellectual property rights. Unlike intellectual property rights, there are no in-built limits to GURTs. Unless the breeder has the biotechnological capability to reverse engineer the GURT variety, the trait is not reproducible through conventional breeding technologies. This means that a GURT-protected innovation remains protected indefinitely. Individuals and nations without biotechnology capabilities have only two very stark choices: purchase the technology or live without it. And this choice becomes even starker over time, as other traits and technologies become available that are dependent on the purchase of the first.

Against this rather dark picture of GURT-based technological progress there needs to be placed the potential benefits from the new appropriation system. One of these benefits concerns the anticipated increase in investments in R&D resulting from increased expected appropriability. This is the aspect of GURTs investigated by Srinivasan and Thirtle in Chapter 6. They run a simulation to ascertain the expected differences in rent appropriation resulting from the requirement of annual repurchases (as under GURTs). They find that rent appropriation increases by a factor of approximately six to eight (depending on the rate of obsolescence of the prevailing technology). So the capacity for firms to benefit from their innovations is improved strikingly by the new technology.

Srinivasan and Thirtle also investigate how this appropriation technology has affected rent appropriation and investment patterns in the one area where it is already in use, the modern hybrid crop variety sector. They report from, a study by Fuglie *et al.* (1996), that the primary hybrid crop (maize) in the USA (1975–92) has a price (per hectare) double that of the major non-hybrid (wheat) and a much more rapid rate of price growth. Fuglie estimates that breeders capture about half of the value of yield improvements in maize, but only about a quarter for wheat. Hence the rate of rent appropriation appears to much greater in the case of the one existing genetic use restriction technology (hybridization). More importantly, the Fuglie study also demonstrates a much greater rate of investment in R&D in the hybrid crops. Maize has private plant breeding investments about ten times that in wheat. Hence the technology of enhanced appropriability is encouraging both increased rent appropriation and R&D expenditures.

It is important to note, however, that (despite this much greater level of private investment) the growth in yield for maize is only about 15 per cent greater than it is for wheat. This is indicative of the fact that the public research sector in the USA has placed much greater emphasis over this period on investments in the improvement of non-hybrid crops. This public sector R&D expenditure is the most likely explanation for growth in the non-hybrids having managed to maintain pace with the hybrids. For some complicated set of political reasons, the impact of hybrid technologies in the USA has been more to redirect public investment than to alter the general manner in which agricultural R&D is undertaken in some fundamental manner.

The second potential benefit from movement toward the GURT-based system is its capacity for opening new markets. In the past, there was little private incentive to invest in the development of crop innovations specifically for poorer countries. These were precisely the places where the institutions would be least developed for the enforcement of any resulting plant breeders' rights. They would also be the places where the individual farmers would have least willingness to pay for traits that were not strictly necessary. The credit markets would be undeveloped for funding either the farmers' or the innovators' interests in investing in improvements. For these reasons, not only has the private sector been based in a small part of the developed world, it has also been focused primarily on the resolution of the agricultural problems in that part of the world. Studies of the plant breeding sector within the developing world find that the investments occurring there are occurring almost exclusively within the hybrid sector (for example, Swanson and Goeschl, 1999). This makes sense, because claims in rights over these varieties are not dependent on the availability of governmental institutions or resources for their enforcement. And since these institutional deficiencies are less problematic in the case of hybrid varieties, we would anticipate that the private sector then would be willing to operate in these markets. Then it might be the case that the development of GURTs would not just enhance the overall level of R&D investments, but also cause investments to occur in places where none had occurred before.

Thus an alternative view on the future of plant breeding under GURTs would be that the private sector would expand its level and range of operations immensely in order to include all countries and all crops. Then the future of modern agriculture would consist of local-level plant breeding operations operating within global networks, finely-tuned to local environmental conditions and popular demands but receiving information and innovations from throughout the global network. The industry would then become far more diverse in its operations and offerings, as local investments made it worthwhile to respond to local demands. It would also be more effective in the aggregate, as it became responsive to local conditions. Most importantly, innovations would diffuse almost instantaneously across the globe, applied to the extent that they made sense under the varying conditions, as multinationals rendered national institutions and boundaries irrelevant.

The fundamental difference between these two alternative views lies in the expected impact of enhanced appropriability on global investment patterns (and hence on institutional change). GURTs should have their greatest impact on appropriability in the developing world, and hence the greatest share of investment and change should occur there as well. If those changes occur, then GURTs may have a substantial impact on the way in which plant breeding operations occur globally. However, if rent appropriation increases without a significant alteration in investment patterns, GURTs will primarily have distributional impacts.

The two chapters at the end of Part II attempt to assess the expected impacts of GURTs. Chapter 7, by Swanson and Goeschl, sets out the basic framework for the analysis of the expected distributional impacts of GURTs. This chapter identifies the primary factor determining the expected impact of GURTs as the individual country's biotechnology capabilities, present or incipient. If a country had biotechnology capabilities, it would be expected to benefit directly from enhanced appropriation technologies. It would be able both to receive enhanced rents from its own innovations, and to make use of some share of others' innovations through reverse engineering strategies. The same would be true in the near future of countries with incipient biotechnology capacities.

More interesting is the case of those countries without any current or incipient biotechnological capacities. What would determine the impact of GURTs on these (the vast majority of all) countries? The factors that the authors identify are (a) the proportion of arable land used for GURT crops, (b) the resistance to foreign direct investment and multinational operations, and (c) the history of the country's performance in regard to hybrid crops. The first two factors are relatively straightforward. GURT technologies are only available for certain crops, and so they will have differential impacts in respect to land area affected. More interestingly, a country's willingness to allow private multinationals to operate its plant breeding sector may be an important determinant of its capacity to benefit from the change to GURTs. Attempts to resist change may only result in placing the individual country at a substantial comparative disadvantage.

A look at the individual country's experience with the most widely-used hybrid crop (maize) is instructive on how widely the experiences might vary. The 'technological gap' between the frontier states and individual developing countries can vary widely, from a −2 per cent yield gap in Egypt to a −96 per cent yield gap in (neighbouring) Sudan. These gaps are representative of the aggregate impact from the combined set of institutional frictions that keep technological advance from diffusing into that particular country. The case of maize provides 40 years of experience with a use restriction technology, experience that is indicative of what to expect to occur in individual developing countries if the technology is expanded.

The final chapter in this section makes use of this insight to present a forecast on the impacts of GURTs on developing countries. It models diffusion of innovation as a catching-up process, looking at how quickly developing countries converge to the frontier states. This assumes that the innovations come from the frontier, but the developing countries are able to absorb and to make use of these innovations with the passage of enough time. The extent to which convergence occurs over a given period of time is a function of the frictions that inhibit it. The study is looking to ascertain whether the use restriction technology (hybrid maize) generated more or less friction to diffusion of innovation. The comparison of hybrid and non-hybrid crop yields across time indicates that there is a significantly reduced rate of diffusion with the introduction of the use restriction technology. In aggregate, the reduced rate of convergence exhibited a difference of about 7 per cent between the hybrid and non-hybrid crop varieties. Developing countries closed the gap in yields by about 31 per cent in non-hybrid crops, but only by about 24 per cent in hybrid crops.

This study does not necessarily demonstrate that the use restriction technology positively inhibits the flow of information across countries, but it does demonstrate that there has not been a substantial reduction in the restrictions by reason of this technological change. This is evidence for the proposition that the enhanced appropriability associated with hybrid crop varieties has not resulted in substantially altered investment patterns. When producers do not respond to changed rent appropriation with altered investment patterns, enhanced appropriation will result more in distributional changes than in efficiency changes.

Thus, Part II demonstrates that the expected impact of GURTs will depend greatly on how the producer sector responds to enhanced rent appropriation. If it invests in a manner that enhances diffusion, the technological change may result in increased innovation and its rapid diffusion worldwide. This is because the multinational firm is uniquely situated to provide easy communication and rapid transfer of technology across boundaries. If the firms do not invest in diffusion, then it is the brute nature of use restriction technologies to inhibit the free flow of information. But it is important to emphasize that it is the combination of use restriction technologies with unaltered investment patterns that will produce the restricted flows of information.

The case study comparing hybrid and non-hybrid crop varieties indicates that the producers have not responded in that case to enhanced appropriation with altered investment patterns. The 40 years of hybrids demonstrate that the developing countries continue to lag in their yields in these varieties. One part of the explanation clearly is that the transfer in technology to developing countries has actually declined in those crops where the private sector is most responsible. This is an unfavourable outcome for developing countries, and it indicates that the shift towards a private sector-dominated plant breeding industry might have important distributional implications.

THE IMPACT ON DEVELOPING COUNTRIES: GENETIC RESOURCES

The final part of the volume concerns the impact of new technologies on developing countries, in regard to their incentives to conserve genetic resources for use in research and development. One of the fundamental aspects of changes in appropriability concerns the distribution of rents between various contributors to yield development. Another aspect concerns the increased efficiency in the use of genetic resources. The chapters in Part III address both issues.

As mentioned much earlier, the production of a plant variety requires inputs not only from the plant breeder but also from many other sources. Many of the plant genetic materials used in producing new plant varieties come from developing countries. This has long been the case. These sources have been increasingly eroded by reason of conversions away from traditional agricultural practices. This means that there is greater value attributed to the still-existing variety in use.

The manner in which rents are appropriated matters for purposes of genetic resource conservation because it is usually the case that rents are appropriable only at one point in the industry. In the R&D process, there are numerous factors combined to produce the information that becomes an innovative plant variety (genetic resources, breeding plots, plant breeders) yet the exclusive marketing right inheres at only one level. In the case of the innovative plant variety, the exclusive right is given to the plant breeder, and it is important that the plant breeder allocate that return efficiently between the various factors of production. If not, there are reduced incentives to conserve and to supply the factors that are required for supplying the innovative plant variety (Swanson and Goeschl, 1999).

The users of plant varieties are also often the suppliers of genetic resources, and so the way in which rents are appropriated is important in determining how they benefit. It is also important in determining how they will supply future plant varieties. GURTs are a mechanism for determining future rent appropriation, and so they might also determine how future resources are conserved and supplied. The chapters in this part survey some of these issues, and indicate how different countries and institutions contribute to this situation.

CONCLUSION

The implication of new technologies for world agriculture is usually a positive, progressive event. The advent of GURTs should herald the same sort of change for agriculture. GURTs represent a new technological opportunity to target research and development to users' needs, and to channel these needs into new investments. It is an opportunity that should not be taken lightly.

Nevertheless, most indicators suggest that this technological opportunity may not ever reach fruition. First of all, this is because the industry itself does not seem to make use of the opportunity to the extent possible. GURTs suggest, not just enhance appropriation, they indicate the possibility of a completely overhauled system of agricultural R&D. In the past it was necessary to have substantial government involvement, in order to advance and diffuse technology. In the future, with GURTs, it might be possible to have both expansive and extensive private sector involvement in the advancement and diffusion of technology.

If this were to happen, the changes inherent in GURTs would represent a revolution in the way in which agricultural development occurred. The private sector would be responsible, not only for advancing the technological frontier, but also for diffusing that technological advance across the whole of the globe. Since there are as many benefits in diffusion as in advance, it would be expected that the private sector would be equally capable of achieving either function. In addition, the participation of the private sector in diffusion would be an important step towards the elimination of unnecessary national restrictions and the enhancement of diffusion.

However, GURTs are not likely to achieve any of this promise. The explanation for this pessimism lies in the case history of the hybrid variety phenomenon. The hybrid varieties represent an early example of a use restriction technology, and they provide over 40 years of experience in the use of such new technologies. The experience from such varieties is not favourable. The history of hybrid technologies is one of enhanced rent appropriation but little change in investment patterns. This implies that the new technology has changed rent appropriation substantially, but efficiency insignificantly. The primary implications have been distributional. Developing countries have seen the benefits from these new technologies, in terms of the diffusion of innovation, diffuse even more slowly than those of non-hybrids. This means that enhanced rent appropriation is changing, not the diffusion of innovation, but mainly the distribution of rents.

There are many policy responses to such an eventuality, ranging from public investments in diffusion to required licensing of new technologies. In sum, however, the primary reason to move towards GURTs is to enable the private sector to manage both innovation and diffusion of technological change in agriculture. It seems counterproductive both to enable this private sector-based management structure and to subsidize the maintenance of a dual structure to run in tandem with it. In the conclusion to this volume we discuss various ways in which diffusion might be aided, if technological change were managed under GURTs.

It is probably a good idea to attempt to aid the private sector in its diffusion of technological change; however, it is probably a better idea to attempt to ascertain what has gone wrong with hybrid varieties. Why has the private sector failed to develop breeding facilities throughout the world? Why do not innovations cross borders easily? What frictions exist that prevent multinationals from moving innovations quickly and inexpensively around the world? These are the questions that should be addressed in the context of hybrid maize. Their resolution will aid significantly the understanding of how GURTs might be made into a beneficial technological change.

REFERENCES

Evenson, R., D. Gollin and V. Santaniello (eds) (1998), *Agricultural Values of Genetic Resources*, London: CABI.

Goeschl, T. and T. Swanson (2001), 'The Social Value of Biodiversity in R&D', *Environmental and Resource Economics*.

Hayami, Y. and V. Ruttan (1985), *Agricultural Development: An International Perspective*, Baltimore: Johns Hopkins.

Juma, C. (1988), *The Gene Hunters*, London: Zed Books.

Swanson, T. (ed.) (2001), *Economics of Managing Biotechnology*, Dordrecht: Kluwer.

Swanson, T. and T. Goeschl (1999), 'Property Rights Issues Regarding Plant Genetic Resources', *Ecological Economics*.

PART I

Setting the Scene: the Framework for Considering Biotechnology's Impacts

2. Population growth and agricultural intensification in developing countries

Nadia Cuffaro

POPULATION AND AGRICULTURAL INTENSIFICATION: CHALLENGES AND SUCCESSES

Rapid population growth has been a feature of modern economic growth. The acceleration following the industrial revolution was very great, but that which has occurred since 1950, mainly concentrated in the developing regions, translated into 'numbers' that are incomparably higher. Figure 2.1 gives a broad idea of such acceleration. Although the world's population growth rate reached a peak in the late 1960s and has declined since, annual

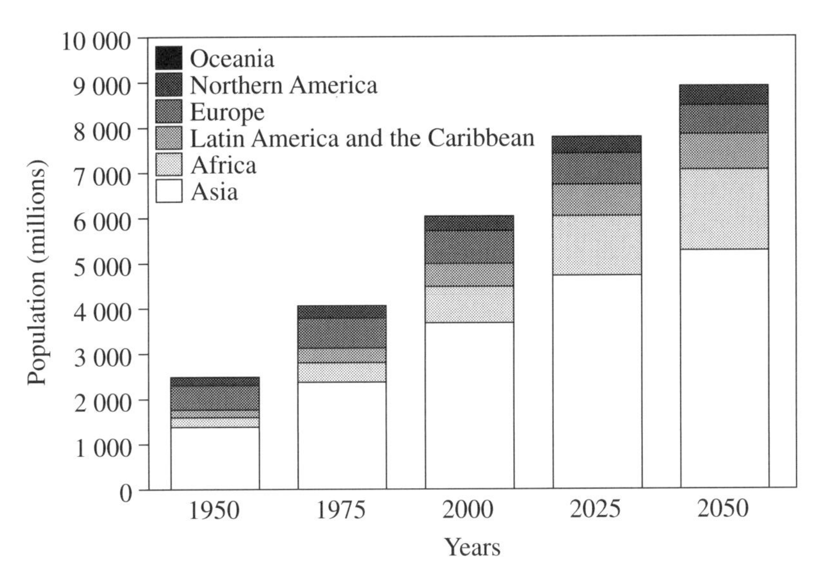

Figure 2.1 Global population growth

population additions have continued to rise because such growth rates are applied to rapidly expanding totals.

Since Malthus, food availability has been a core argument of the population scare. The basis of Malthus's pessimism was the idea that population growth would translate into increasing pressure on land and therefore, in the absence of technical progress, into decreasing returns to labour. If population growth is associated with decreasing land to labour ratios, a possible pessimistic scenario – outlined by Khan (1991), who refers to the experience of Bangladesh in the 1950s – entails not just lower agricultural incomes, but also landlessness. The resulting wage-labour supply increase and the increased factor share of land then translate into lower real wages.

Population growth may not imply the lowering of land to labour ratios if the land frontier can be expanded or if labour absorption in the non-agricultural sector is fast (or substantial migration outlets exist, as was the case for modern European growth). Both possibilities are severely limited for most of today's developing countries and in most of them population pressure on land has increased since the 1960s. Yet, in contrast to the Malthusian prediction, production in the developing countries has kept pace with population: per capita agricultural production has increased during the last three decades, on a world scale and in all developing regions with the exception of sub-Saharan Africa. Such growth has occurred in the context of a long-term decline in the real price of food.

These trends can be taken as prima facie evidence of the ability of agriculture to respond to increasing population without incurring Malthusian outcomes, and they are the result of the variable which was missing in Malthus's reasoning: technical progress.

In Malthus's model food production establishes the limits to population growth. Boserup (1965), however, has provided rich evidence to support a model of endogenous technical progress in agriculture, where increased population pressure on land leads to shorter fallow periods (from forest–fallow cultivation to multicropping) and to corresponding changes in agricultural methods and tools. Growing population implies increasing labour/land ratios with associated diminishing returns until, eventually, a new, superior technique is introduced.

Boserup's argument is twofold: increasing use of labour along a given production function lowers average product, and may bring it close to the subsistence level; new, more intensive technologies become superior (that is, they produce higher yields per acre) only as density increases, but they may not translate into higher labour productivity and may require more working hours per worker. Therefore cultivators will choose more intensive methods only when a certain density of population has been reached and the supply of food becomes tight.

Many theoretical shortcomings limit the validity of the Boserup model. The egalitarian structure of the model, whereby increasing food needs induce technological responses and increased food production caters to the needs of the community, overlooks all the questions of distribution and entitlement;[1] the model focuses on technology transmission – it does not explain the entire process of technical change, which includes research, discovery and adoption; institutional adaptation (in this case the research system and property rights on land) is not explained; and the focus is on low-input agriculture, neglecting the fact that limited availability of capital to acquire inputs may be a constraint on technology adoption. Finally, and perhaps crucially, the model is based on historical data and slow processes of adaptation, and neglects some possible critical implications of the rapidity of changes and/or the already high densities in modern poor countries.

Nevertheless, several country studies have confirmed Boserup's main arguments (Cuffaro, 1997) and it is difficult to disagree with Lipton's (1990: 223) observation that the Boserup model, unlike that of Malthus, may offer an explanation for the difference in vital statistics and agricultural growth performance between land-abundant Africa and South and East Asia.

Optimism over the questions left open by Boserup is suggested by the Hayami and Ruttan (1985) induced innovation model (IIM). The IIM (Figure 2.2) provides an economic theory of invention and adoption of

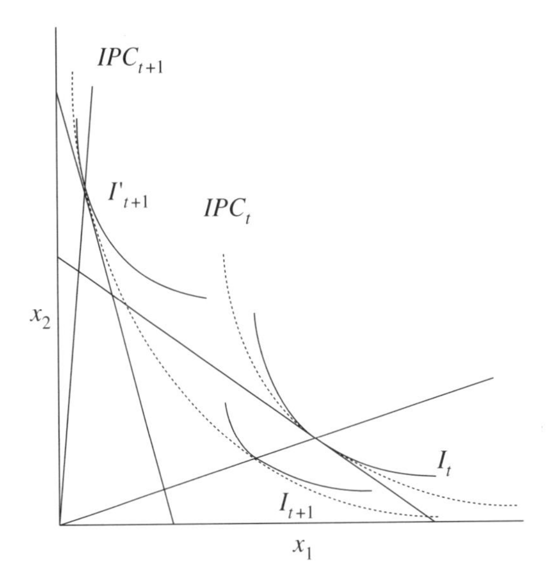

Figure 2.2 The induced innovation model

technologies, whereby relative price changes determine responses by economic agents and research institutions, guiding technological progress towards saving the factors that become relatively scarce. Thus, as population grows, land prices increase relative to wages and to the prices of man-made, land-saving inputs (such as fertilizers). Technological change therefore moves in a land-saving, labour-using direction. Moreover, Hayami and Ruttan argue that changes in underlying economic conditions (increasing population pressure on land, increasing product demand because of trade opportunities, and technical change) also induce the development of property rights and contractual arrangements that allow a more efficient allocation of resources.

In this framework shifts in the stock of knowledge (represented by the innovation possibility curve, IPC) are neutral (factor proportions would remain the same if the same price ratio prevailed at time $t+1$), whilst relative factor prices determine the factor bias of the techniques invented. The 'green revolution' can be seen as the switch to a technology (I'_{t+1}) using relatively less land (x_1) and more fertilizer (x_2) (in conjunction with land infrastructure) as compared to *It*, in response to a decrease in the relative price of fertilizer. The distance between IPC_t and IPC_{t+1} represents the technology gap between developed and developing countries; the IPC_{t+1} became available to the national research systems in the developing countries through the transfer of crop varieties developed by international centres, publicly financed.

Historically, the green revolution is the result of a number of developments both in the demand for innovations (in developing countries) and in the supply of innovations (mainly in the developed countries) which accelerated after the Second World War (Hayami and Ruttan, 1985; Hayami and Otsuka, 1994).

On the demand side, many developing countries, especially in Asia, were experiencing rapid population growth and a substantial exhaustion of the land frontier, that is, of the possibility to expand cultivation to new land. On the supply side, there existed a large accumulated gap in the application of science to agricultural production between the developed and the developing countries. Furthermore, the relative price of fertilizer had declined as a result, first, of increased productivity in the fertilizer industry in developed countries, and later of the growth of domestic fertilizer production in the developing countries through technology transfer. Within this framework, it is understandable why the green revolution succeeded mainly in Asia, where population pressure on land, and therefore presumably the rice/fertilizer ratio, were higher (Hayami and Ruttan, 1985; Hayami and Otsuka, 1994).

Food production in Asia has more than kept pace with demand over the past 30 years and in many areas grain yields are among the highest ever achieved. Without the green revolution, such supply increase would not have been possible and supply expansion would have occurred at the cost of opening new lands and depleting the soils. The sharp increase in the rate of productivity growth linked to the initial replacement of traditional with modern varieties (MVs) has been followed by lesser but continuous increases (with newer generations of MVs periodically replacing the original MVs) whose cumulative effect is substantial. Furthermore, there has been a trend towards diffusion of MVs in rainfed areas – from wetter to drier – and research, especially in wheat and, to a lesser extent maize, has also increasingly addressed the problems of pest and pesticide resistance and nutrient efficiency (Byerlee, 1996).

In summary, while agriculture was the traditional (Malthusian) basis for population pessimism, within the more recent debate on the linkages between population growth and economic development, references to agriculture are often used as examples of the ability of societies to produce the right economic and institutional responses to demographic shifts. Such a view is supported by the trends in per capita production and prices and by the models of endogenous, population-driven, technical progress and of induced, factor-saving innovations (as well as by the theory of the evolution of property rights on land) and by a related body of empirical evidence on contemporary less developed countries.

By and large, the theoretical background of the analyses that suggest population optimism for agriculture lies in a more general belief that the power of the invisible hand works also for the institutions: population growth, by changing certain relative prices, would produce responses – chiefly in terms of technical progress and property rights – that essentially take care of Malthusian (and of traditional neoclassical) concerns about population.

POPULATION AND AGRICULTURAL INTENSIFICATION: FAILURES

The global trends in production and prices do not demonstrate the ability of agriculture to match the growth of effective demand without harming the environment and its own resource base, nor do they exclude the possibility of severe local failures. Furthermore, millions of people in the developing countries do not have access to enough food because they are too poor to demand and/or to produce more food (FAO, 2000)

While the review of past experience and of theories basically confirms that population growth triggers adjustments in agriculture in terms of intensification, it also reveals that adaptation processes are complex and difficulties are more likely to arise when population growth rates are very fast and/or densities are already high in the context of poverty and/or unfavourable natural conditions (Cuffaro, 1997; Cuffaro and Heins, 1998). The technological and institutional adjustments required by population growth often have not emerged at all, or not emerged on time; that is, the broad historical validity of the relationships postulated by Boserup and the possibility of the adjustments foreseen by the IIM do not mean that under all conditions one may expect a smooth transition from low to high densities, together with sustainable intensification.

With rapid population growth, in the context of poverty and unfavourable climate, communities have often failed to achieve the needed intensification, or have proceeded too far along a given production function, causing a reduction of long-term land productivity, before a new technology was introduced. There is a strong relationship between poverty and possible failures of adjustment to population growth. One important reason is remarked by Lipton (1990). Addressing the link between necessity and invention, which is at the basis of the Boserup model of induced innovation, Lipton points out that necessity must not be hunger alone, but rather hunger backed by effective demand, and that this may not be the case in very poor, intensive agriculture regions, with severe income inequity.

In general, sustainable output may be lower than the maximum attainable in the short run, while farmers or pastoralists may not be able to postpone production if they are merely achieving subsistence. Poverty also reduces the options available to farmers in terms of substitution of man-made inputs for natural resources and in terms of investments in resources maintenance. Welfare-poor households may also be 'conservation-investment' poor; that is, they may lack the ability to make the investments required to maintain or enhance the resource base, and to prevent or reverse degradation (Reardon and Vosti, 1997). Dasgupta (1993) has actually hypothesized a self-reinforcing mechanism between environmental degradation, poverty and high fertility rates, partly confirmed by the empirical evidence of a World Bank (1991) study on sub-Saharan Africa.

Formal research and extension systems hardly have a role in Boserup's analysis, while Hayami and Ruttan explicitly claim that such systems will (eventually) provide the right responses to changing factors' scarcity. However, national research systems may encounter significant difficulties in responding to rapid changes. For example, it is widely held that the response of research institutions to rapidly growing population in most of Africa has been inadequate. The reasons for this are three sets of factors affecting the performance of the national agricultural research systems. The first is the general low stage of development of institutions at independence, including the low initial conditions in terms of scientific and education base, and the fact that the attainment of independence is relatively recent. This had far-reaching consequences for the research system. Second is the nature of the African environment and of its pattern of agriculture: a range of staples is grown across the continent and even at the single farm level, in contrast with Asian dependence on a single crop. Third is the fact that both the scientific paradigm and the varietal development that characterized the green revolution were far better suited to Asian, rather than African, conditions (Eicher, 1990; Lynam and Blaikie, 1994; Pardley *et al.*, 1997; Rukuni *et al.*, 1997).

On the whole, the performance of agricultural research in Africa illustrates a 'timing' problem: Africa has experienced an extremely rapid process of transition from land abundance to increased population pressure on land, in the context of difficult natural environments. Society's response in terms of research could not be fast enough. Even if adaptation does occur when the underlying economic conditions change, the starting conditions and the pace at which such changes occur are critical. It may thus happen that, although communities try to adapt, the rapidity of change, coupled with various constraints, leads in many cases to a decline in long-term land productivity.

The limits of institutional functionalism are particularly strong in the context of an analysis centred on population. The institutional analysis usually recognizes that many factors may lead to suboptimal outcomes and/or slow down the adjustment pace. But even in the best of all possible worlds population-induced adjustments in the institutional sphere, where slowly changing cultural factors have a large role, may lag behind the type of fast population growth experienced by modern less developed countries. Poverty, difficult natural environments and inequality in the structure of asset ownership make adjustments more difficult and interact negatively with the 'time' factor.

The latter issue can also be illustrated with reference to the green revolution type of technological change. This was obviously more complicated than Figure 2.2 shows, and its preconditions were also more complex than the change of the price ratio which triggers a technological response in Figure 2.2. As Hayami and Ruttan recognize,

> Adjustments . . . usually involve time and costs. The development of fertilizer-responsive MVs requires investments in research. Better husbandry practices must be developed and learned. Complementary investment in irrigation and drainage may be required to secure adequate control of water. It takes time to reorient the efforts of public agencies in such directions in response to price changes. It is particularly costly and time consuming to build adequate institutions and competent research staff. (Hayami and Ruttan, 1985:276)

Hence there are two sets of related problems: one is a problem of sequencing, the other is a problem of institutional responses. Lipton observes that technological change

> cannot, economically, turn 'this Island into a garden' in any sequence one wishes. The appropriate sequences usually require both public and private investments substantially higher than are likely. . . . With some exceptions, biological, chemical, agronomic and mechanical innovations – in the absence of improved structures for water management – have, since 1960, proved unable to support profitable farmers' decisions that permit yields to keep up with, let alone outpace, rural population growth in developing countries. But, given water control, green revolution-type responses to population growth have been rapid in many years. (Lipton, 1997:85)

Hence an additional reason for the success of the green revolution in Asia is the fact that historically high population pressure led to agricultural intensification and to the creation of an irrigation infrastructure which, together with farmers' experience in irrigated agriculture, established an essential precondition for the diffusion of the new varieties. By the same token, the large difference in rates of adoption among regions in Asia would be explained by the extent of water control (Hayami and Otsuka, 1994).

Hayami and Ruttan (1985) point out the critical role of the relative pace of population growth and institutional adjustment. Communities first adapt to population growth by opening new lands but, as inferior lands are reached, irrigation becomes a more profitable base for agricultural growth. In turn this infrastructural development enables the diffusion of the new seed-fertilizer technology. However, there can be a substantial time lag in adjusting to the new economic conditions. Since irrigation systems tend to have a number of public goods characteristics, group action by farmers for local systems and public investments for large irrigation projects are required.

> such organizational capacity and habit grow in a rural society over time, perhaps several generations . . . the government, of course, may fill this gap. But the allocation of public resources is also a public process involving compromises among vested interests, therefore it is unlikely that government investments in irrigation will provide an immediate response to the changes in man/land ratios. (Hayami and Ruttan, 1985:312)

Furthermore, the likelihood of successful collective action in irrigation and drainage partly depends on the structure of asset ownership. An unequal structure of land ownership in agriculture will represent an obstacle to the development of the irrigation infrastructure. On the one hand, small farm size and fragmented holdings make it uneconomical for most farmers to install pumps individually; on the other hand, group action by farm producers – who could undertake irrigation projects using seasonally idle labour – is hampered by the inequality in the distribution of land assets (Hayami and Ruttan, 1985). A similar argument has been made by Boyce (1987) with reference to the difficulty of collective action for irrigation projects in Bangladesh.

Other authors have shown that there is a relationship between the asset ownership pattern and the success of the green revolution: small peasant farming based on family labour – which is more dominant in the food crop sector in Asia, as compared to Africa and Latin America – is much better suited to the modern technology than farming based on hired labour. Both de Janvry (1973) and Boyce (1987) have shown, respectively in the cases of Argentina and Bangladesh, that the time lag in the induced innovation mechanism, linked to an unequal structure of asset ownership in agriculture, can be substantial, and can therefore depress the long-run growth rate of agriculture and of the standard of living.

In summary, the success of the green revolution was also linked to a set of institutional factors and, although there has been a trend towards diffusion of MVs in rainfed areas, MVs of rice have not been adopted widely in areas where water control remains very poor (and, in general, MVs have not been successful in areas where drought stress is frequent) Byerlee (1996).

FUTURE TRENDS AND CHALLENGES

Table 2.1 shows the projected growth rates of population (UN medium variant), aggregate demand and production as estimated by FAO (2000). The difference between the past growth rates of demand and FAO projections is nearly equal to that of the population growth rates. The deceleration is essentially explained by trends in a group of developing countries, including many of the largest in terms of population (China, Indonesia, Brazil, Mexico, Nigeria, Egypt, Iran, Turkey), that started in 1995/7 with fairly high per capita food consumption (over 2700 kcal/person/day) and are experiencing a significant slowdown in their population growth. In contrast, according to FAO, in the other developing countries, including India, demand growth will decelerate less than population.

Hence, a global slowdown of demand for agricultural products will coexist with historically high growth rates in many areas of the world. At the same time, the possibility of expanding cultivation to new lands no longer exists in many areas of the developing world, while in others such

Table 2.1 Projected growth rates of population, agregate demand and production (per cent p.a.)

	1995/7–2015			2015–2030		
	Pop.	Dem.	Prod.	Pop.	Dem.	Prod.
World	1.2	1.6	1.6	0.8	1.3	1.3
Developing countries	1.4	2.2	2.1	1.0	1.7	1.6
Developing countries excl. China	1.6	2.3	2.2	1.2	1.9	1.8
Sub-Saharan Africa	2.4	2.8	2.6	2.0	2.5	2.4
Latin America and Caribbean	1.4	2.0	2.0	0.9	1.5	1.6
Lat. Am. and Car. excl. Brazil	1.5	2.1	2.0	1.0	1.6	1.6
South Asia	1.5	2.6	2.5	1.0	2.1	2.1
East Asia	0.9	1.9	1.8	0.5	1.3	1.2
Industrial countries	0.3	0.6	0.8	0.1	0.4	0.6
Transition economies	0.0	0.9	1.1	−0.1	0.6	0.6

Source: FAO (2000).

expansion could only occur at the expense of forests. Future demand growth will have to be met essentially by increasing the productivity of land already in cultivation. However, on the one hand, unless the promises of biotechnology materialize, the sources of future productivity growth in agriculture seem less discernible and achievable than at the beginning of the green revolution, on the other hand, agricultural intensification, in general and along green revolution lines, has mostly been associated with significant environmental spillovers.

Ruttan observes:

> incremental responses to increases in fertiliser use have declined. Expansion of irrigated area has become more costly. Maintenance research . . . is rising as a share of total research. The institutional capacity to respond to these concerns is limited, even in the countries with the most effective national agricultural research and extension systems. Indeed, during the 1980s, there was considerable difficulty in many developing countries in maintaining the agricultural research capacity that had been established in the 1960s and 1970s. (Ruttan, 1997:24)

The successful, intensive systems based on conventional green revolution technology are experiencing two related problems. Some authors have argued that the yield gap between experiment stations and farms may be closing and that there may be a stagnant or even declining technological yield frontier. Hayami and Otsuka (1994) remark that the higher average rice growth rate registered in South and Southeast Asia in the 1980s, as compared to the 1970s, was due to the expansion of modern varieties to less favourable production environments and to the development and diffusion of varieties more resistant to pests and diseases, but not necessarily higher yielding than earlier MVs under ideal rice growing conditions.

More disturbingly, Pingali (1997) suggests that the technological yield frontier may be declining because of a degrading paddy microenvironment resulting from intensive rice monoculture. Rice production has been subject to increased pest pressure, attributed to the uniformity of varieties grown, indiscriminate pesticide use and increasing susceptibility of varieties to resistance breakdown. Relatively minor pests have caused noticeable losses as the area planted to modern varieties increased. Furthermore, there is evidence of rapid depletion of soil micronutrients and changes in soil chemistry brought about by intensive cropping or increased reliance on low-quality irrigation water. In summary, 'There is a growing concern about the resilience of the humid tropical lowlands and an increasing understanding that they do not have an unlimited absorptive capacity' (Pingali, 1997:225–6).

More generally, intensive agricultural systems have well known and significant environmental spillovers. While some technological developments, like the development and diffusion of hybrid rice varieties, suggest the possibility of moving further with fertilizer- (and other chemical inputs-) intensive technology, the external diseconomies arising from the use of such inputs are increasingly recognized (Hayami and Otsuka, 1994).

BIOTECHNOLOGY

In principle, biotechnology could provide answers to many of the problems discussed in the previous sections. Innovations, such as the development of an insect- or herbicide-resistant plant variety,[2] can be represented as a non-neutral shift of the technological frontier. In the case of *Bt* varieties (Figure 2.3) the technology replaces a conventional chemical input (insecticide) with seeds incorporating insect resistance.

In the case of *RR* varieties (Figure 2.4), seeds incorporating resistance to *Roundup* herbicide substitute for labour and other herbicide management inputs when used in combination with *Roundup*.

These shifts are produced by the combination of two developments: rapid advances in molecular biology and genetic engineering and the establishment of intellectual property rights (IPR) on innovations incorporated into seeds.[3] Firms carry R&D and offer the new products because with IPR protection they can appropriate the benefits. The new varieties change the structure of costs for farmers: they reduce some variable costs (such as insecticides costs) and increase seed costs; farmers in the developed and

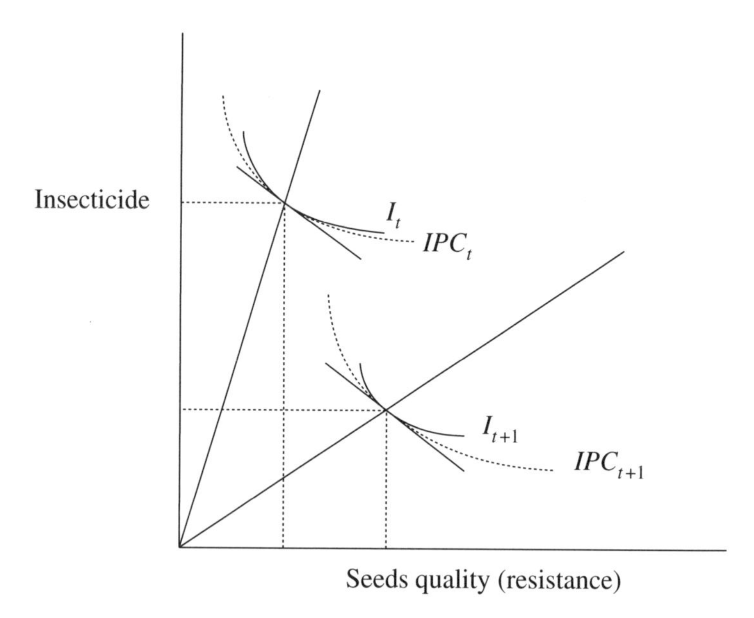

Figure 2.3 Bt varieties

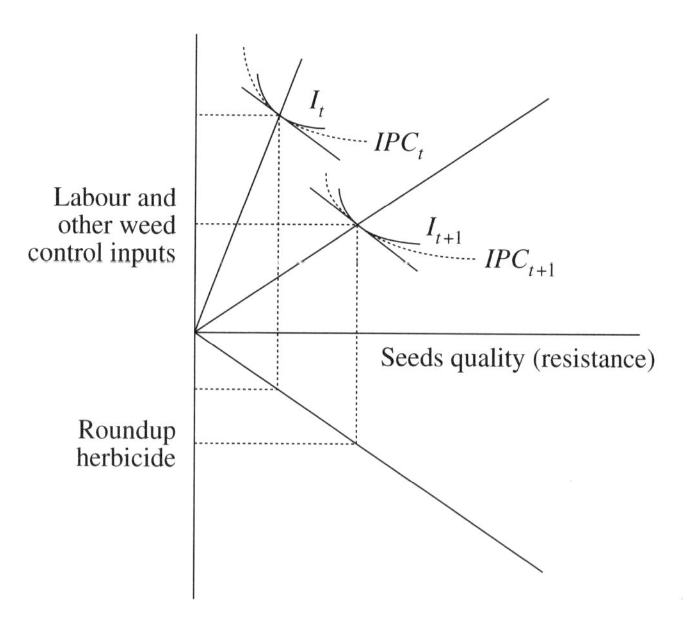

Figure 2.4 RR varieties

developing countries adopt the new technology if, on the whole, it is cost reducing and pay a price premium for the modified seeds.

Biotechnology does substantially reopen the technology gap in agricultural innovations between the developed and the developing countries. As compared to traditional plant breeding, it allows responses that are faster (and the importance of 'time' has been extensively discussed in the previous sections), more flexible (that is, the innovations can fit the needs of many different environments, and therefore do not require the set of conditions that limited the success of green revolution varieties to certain regions of the world), more precise (that is, they produce varieties whose traits correspond more precisely to the research objectives than those obtained through conventional plant breeding) and have the potential to address some of the externality problems of agricultural intensification.

Plant breeding has evolved over time, exhibiting a dramatic increase in the speed with which new varieties are produced, from the several generations required for developing a new variety in the context of informal innovation by farmers to the great acceleration achieved by scientific plant breeding during the last century. Modern biotechnology represents a quantum leap in this process. For example, molecular marker technology can halve the time needed to produce new varieties with resistance to important crop diseases. The same technique accelerates developments as varied as pest control traits (like genetic resistance to insects); agronomic traits such as tolerance to drought, salinity and heat; post-harvest traits such as delayed ripening of fruits; and output traits such as specific food qualities. Transgenic approaches considerably broaden the range of gene pools that can be used for crop improvement purposes, as they allow the introduction into crop plants of genes from any class of organisms (Maredia *et al.*, 1999; Lindner, 1997).

As for the issues that have dominated the public debate, it should be recalled that the possible negative side effects of biotechnology,[4] are in many ways analogous to the type of externalities associated with intensive agriculture, and it is not clear whether, on balance, problems such as pest resistance and loss of biodiversity may be reduced or amplified with the new technologies. However, while the negative externalities of agricultural practices such as chemical inputs use are known, the side-effects of the genetically modified (GM) crops are surrounded by a higher degree of uncertainty, because biotechnology allows genetic crossing among very distant classes of organisms. Uncertainty and the related consumers' diffidence raise the institutional costs of innovations both for the private and for the public sector since, for instance, the safety assessment on GM crops must be more careful than with the outputs of conventional plant breeding (Dale, 2000).

Given the great potential of biotechnology, one could imagine seeds that provide most of the answers to the problems of poor farmers (low-input agriculture) in unfavourable environments and therefore to most of the problems of intensification discussed above. These seeds would combine increased resistance to pests and to environmental stresses such as drought, and improved nutritional quality. Furthermore, the embodiment of desirable qualities in seeds reduces both the human capital requirements of farming at the farm level and the dependence on input markets (other than the seed market), both desirable qualities. In practice, however, the impact of biotechnology on the problems of agricultural intensification discussed in the preceding sections is largely dependent on the evolution of a number of market structure and institutional issues.

Biotechnology is closely linked to the establishment of intellectual property rights in agricultural innovations, in contrast to the tradition of sharing of genetic materials between public research centres engaged in plant breeding. The establishment of IPR should foster innovation to the extent that it provides private incentives for R&D, and the historical experience of hybrid maize[5] in the USA, and other more recent examples, actually show that IPR definition is associated with very high rates of private R&D investments and of varietal development. If biotechnology translated solely into greater private investments it would unambiguously contribute to growth. However, past experience shows that IPR definition resulted in a 'crowding out' of public spending in plant breeding; furthermore, biotechnology currently has been associated with industry concentration.

The biotechnology sector has rapidly moved from a competitive structure with a large number of small innovative firms to an industry dominated by six large multinational chemical and pharmaceutical corporations (Monsanto, for instance, was a traditional chemical company) through a process of mergers, acquisitions and joint ventures. One possible set of explanations for this trend is related to the very process of IPR definition: both to its nature and to its still unsettled status. The definition of intellectual property rights on innovations typically tries to achieve the right balance between providing incentive to discoveries (granting the innovator the right to prevent their use by others without permission) and ensuring the technology transfer (for example, by limiting in various ways the breadth of patents and imposing that the invention be described in a way that permits duplication on patent expiry).

However, the variety and reach of IPR claims related to biotechnology processes and inventions, including those that represent enabling technologies for research, has a number of consequences. First, it is a direct source of monopoly power. Second, it entails a substantial amount of litigation (that is, substantial transaction costs). This in turn is an incentive to concentration. Firms would acquire related companies in an effort to reduce contractual hazards, by internalizing IPR disputes. Third, the 'gridlock' of patent rights could slow down the pace of innovations, although the biotechnology companies try to ensure their 'freedom to operate' (even while engaging in litigation in the courts) by reaching explicit or tacit agreements on cross-licensing (Lindner, 1997). Finally, the necessity to seek and defend intellectual property rights over key technologies and a complex system of regulations impose high fixed institutional costs, which represent a further incentive to concentration.

There are fears that lack of competitive pressure may eventually result in a reduction of the pace at which innovation proceeds. As for the role of the public sector, the experience with hybrid maize in the USA shows that the protection of property rights results in a very high rate of investment and variety development and in the dominance of private seed companies. The replacement of public by private plant breeders is an indication of the latter having a competitive advantage over the former. However, in the case of biotechnology, there are also indications that the public sector may have some disadvantages vis-à-vis large biotechnology companies in defending its 'freedom to operate'. Tentative explanations include the fact that its internal decision mechanisms make it costly to negotiate cross-licensing and other agreements, and the fact that public institution would not sue for anti-competitive behaviour (Lindner, 1997).

Furthermore, the public research centres have experienced for decades the consequences of financial restraint and have been under pressure to become partially self-financed and to replace public with industry-provided funds. Biotechnology has increased this pressure.[6] On the whole, it seems clear that, if the public sector is to retain a distinct role in biotechnology, a substantial institutional response is required at the national level and in the international institutions.

From the developing countries' point of view, there are a number of issues. The research priorities are increasingly defined by the private life science companies in the developed countries. The private sector is guided by the profit motive and therefore by the extent of the effective demand for innovations, that is, by the extent of the actual or potential value of the market for seeds. Therefore research includes all the major food crops;[7] the life science companies should be expected to try to introduce available products (innovations) in many developing countries and to address directly specific developing countries' problems that involve large numbers of non-poor farmers.[8]

However, the discussion of agricultural intensification in response to population growth has indicated that most of the historical failures have been related to poverty and/or very unfavourable environments. The low input–low yield equilibrium of smallholder/peasant farming systems in many parts of the developing world is the result of a number of constraints: cash and credit constraints limit the possibility of acquiring inputs; poor infrastructure in terms of input supplier networks means that input deliveries may be late and inputs may be of poor quality;[9] poor infrastructure in terms of storage and roads means that output price fluctuation tend to be wide locally.

Farmers in these systems use carry-over planting material. With biotechnology they should eventually buy seeds but this may happen only if those seeds allow them to overcome some of their economic and natural constraints. The experience with hybrid maize – which requires all the inputs (purchased seeds, fertilizer, pesticides and good rain) at the appropriate time – shows that an innovation which does not address those constraints has a low probability of being adopted. The public research systems in the developing countries and the international research system have a role in addressing the demand for innovations of poor farmers and the safety concerns of consumers in the 'backward' countries.

With biotechnology the increasingly private good nature of innovations will not allow the type of transfer that has taken place with the green revolution; that is, it cannot be assumed, as in the induced innovation model, that the IPC will be the same worldwide, or that the time lag with which the 'frontier' technologies become available in the 'backward' countries becomes longer. In the meantime, adopting farmers in the 'backward' country pay rents to IPRs that belong to foreign seed companies.

With the exception of a few large countries (including India, China, Mexico and Brazil), developing countries have little research capacity and no regulatory framework to acquire legally and release safely biotechnology products (Maredia *et al.*, 1999). Furthermore it appears that the role of the private sector in terms of biotechnology R&D is still very limited. Maredia *et al.* list two reasons why biotechnology actually poses problems of public investment decisions in the developing countries. One is the opportunity to exploit the technology spillovers – from the developed countries' R&D sector – that may easily occur at the process level. The other is the difficulty of developing a substantial private involvement in biotechnology R&D in countries where the size of the market is small (because farmers are poor) and/or IPR enforcement too expensive (because small farms are prevalent). They hold that, 'For most self-pollinated crops in small-farm agriculture, the public sector, both international or national, will continue to play the leading role' (ibid.:7).

Biotechnology requires at least some type of investment in high-skills human capital (in the biological and legal fields) and regulatory frameworks. One may argue that some of these adjustments are no more challenging than those required by the green revolution. For example, it is no less difficult to develop water control in a vast countryside, involving a large number of farmers and local institutions, than to 'produce' PhD students and regulatory frameworks. Both can be obtained on the basis of the institutional build-up in the developed countries, presumably at very low cost. On the other hand, there may be costs associated with the fact of being entirely 'policy takers' in terms of regulatory frameworks for IPRs and safety and the implementation of such frameworks may be difficult.

CONCLUSIONS

Although the growth rates of population and demand for food will continue to slow down, the capability to satisfy the growth of effective demand in a sustainable manner presents a formidable challenge for agriculture. Biotechnology has an enormous potential but poses a set of issues of regulation and institution building, both in the developed and in the developing countries. In the latter, investments needed to build even a minimum level of biotechnology research capacity include investments in research capacity, in the capacity to review and manage the environmental and human safety implications of GM crops, and in a regulatory framework for IPR protection. Although the challenge is formidable, one must recall that there may be few technological alternatives and also that the success of the green revolution was linked to a set of institutional preconditions.

NOTES

1. This includes the fact that, if population growth is associated with increasing landlessness, technical change must not only make it possible to increase land productivity: it must also be labour-using, or an increasing number of unemployed wage workers may be exposed to starvation.
2. Herbicide-tolerant and insect-resistant plants are the most widely used agricultural biotechnology innovations. *Bt* varieties of maize and cotton incorporate a genetic sequence of a microorganism (*Bacillus thuringiensis*) so that plants produce proteins toxic to certain species of insects. Monsanto's *RR* (*Roundup-Ready*) soybeans incorporate genes that permit the plant to tolerate a very effective herbicide (*Roundup*) previously developed by Monsanto.
3. With the significant exception of hybrid corn in the USA, farmers' free access to plant varieties has historically been the rule in agriculture. In the USA, although legislation protecting plant breeders' IPRs had been introduced in the 1930s, decisive steps in this direction have been linked to modern biotechnology, with the granting of utility patents encompassing living organisms. Furthermore, biotechnology has allowed the development of specific techniques designed to protect property rights. Genetic use restriction techniques, known as 'terminator genes', allow perfect property rights enforcement on proprietary seeds, as plants produce seeds that are sterile and must therefore be purchased each season.

4. In the field, *Bt* varieties could induce the evolution of *Bt*-resistant insects and also kill non-target insects, modifying the ecological equilibrium; insect and herbicide resistance could be transferred to weedy relatives. Globally, it is feared that the role of multinationals in biotechnology innovations may lead a drift towards genetic uniformity and genetic diversity erosion. As for human health, the use of genes for antibiotic resistance in biotechnology raises worries about the possibility that such resistance may be transmitted to consumers of genetically modified organisms (GMOs).
5. Property rights are enforced for hybrids because hybrid seeds, when replanted, lose productivity. Farmers must buy seeds each planting season.
6. In the USA, the role of royalties, grants, contracts and donations from the private sector has substantially increased. Some universities have accepted equity stakes in private companies as payment for the transfer of technology (Zilberman *et al.*, 1999).
7. Hence crops that are specific to relatively small areas or poor agricultural environments (such as cassava and yams) have been termed 'orphan commodities'.
8. Traxler *et al.* (1999) quote as an example the development by Monsanto of a genetically modified variety of *Bt* soybean effective against an important pest in Brazil.
9. Private companies may sell pesticides without standard tests or controls, chemicals are often past their sell-by dates and, in some instances, bootleg (false) compounds have been sold. Farmers have reasons not to trust the private (often newly privatized) companies (Robinson, 2000). In discussing reasons for low adoption of maize hybrids in Ghana, Santaniello (2000) quotes the poor performance of the government-owned seed company in supplying farmers with quality seed in a timely fashion.

REFERENCES

Boserup, E. (1965), *The Conditions of Agricultural Growth*, London: Allen & Unwin.

Boyce, J.K. (1987), *Agrarian Impasse in Bengal. Institutional Constraints to Technological Change*, Oxford: Oxford University Press.

Byerlee, D. (1996), 'Modern Varieties, Productivity and Sustainability: Recent Experience and Emerging Challenges', *World Development*, 24(4), 697–718.

Cuffaro, N. (1997), 'Population Growth and Agriculture in Poor Countries: a Review of Theoretical Issues and Empirical Evidence', *World Development*, 25(7), 1151–63.

Cuffaro N. and F. Heins (1998), 'Population growth, food production and land degradation in poor countries', in J. Restoin (ed.), *Colloque Mondialisation et Géostrategies agroalimentaires*, Montpellier: INRA and World Congress of Environmental and Resource Economists, *http://www.feem.it/gnee*.

Dasgupta, P. (1993), *An Inquiry into Well-Being and Destitution*, Oxford: Clarendon.

Eicher, C. (1990), 'Building African Scientific Capacity for Agricultural Development', *Agricultural Economics*, 4, 117–43.

Evenson R., W. Lesser, V. Santaniello and D. Zilberman (eds) (1999), *The Shape of the Coming Agricultural Biotechnology Transformation: Strategic Investment and Policy Approaches from an Economic Perspective*, CABI.

FAO (2000), *Agriculture: Towards 2015/30*, Technical Interim Report, Rome: FAO.

Hayami, Y. and V. Ruttan, (1985), *Agricultural Development: An International Perspective*, Baltimore: Johns Hopkins University Press.

Hayami, Y. and K. Otsuka (1994), 'Beyond the Green Revolution: Agricultural Development Strategy into the New century', in J.R. Anderson (ed.), *Agricultural technology: policy issues for the international community*, CAB International and the World Bank.

Khan, A. (1991), 'Population Growth and Access to Land: An Asian Perspective', in R. Lee *et. al.* (eds), *Population, Food and Rural Development*, Oxford: Clarendon, pp. 143–61.

Lindner B. (1999), 'Prospects for Public Plant Breeding in a Small Country', in R. Evenson, W. Lesser, V. Santaniello and D. Zilberman (eds).

Lipton, M. (1990) 'Responses to Rural Population Growth: Malthus and the Moderns', in G. McNicoll and M. Caine (eds), *Rural Development and Population, A Supplement to Vol. 15 Population and Development Review*.

Lipton, M. (1997), 'Accelerated resource degradation by agriculture in developing countries? The role of population change and responses to it', in S.A Vosti and T. Reardon (eds), *Sustainability, Growth and Poverty Alleviation*, Baltimore: IFPRI and Johns Hopkins University Press.

Lynam, J.K. and M. Blaikie (1994), 'Building Effective Agricultural Research Capacity: The African Challenge', in J.R Anderson (ed.), *Agricultural technology, policy issues for the international community*, CAB International and the World Bank.

Maredia, M., D. Byerlee and K. Maredia (1999), 'Investment Strategies for Biotechnology in Emerging Research Systems' in R. Evenson, W. Lesser, V. Santaniello and D. Zilberman (eds).

Pardley, P.G., J. Roseboom and N.M. Beintema (1997), 'Investments in African Agricultural Research', *World Development*, 25, 409–23.

Pingali, P. (1997), 'Agriculture–Environment–Poverty Interactions in the Southeast Asia Humid Tropics', in S. Vosti and T. Reardon (eds), *Sustainability, Growth and Poverty Alleviation*, Baltimore: IFPRI and Johns Hopkins University Press,.

Rausser G., S. Scotchmer and L. Simon (1999), 'Intellectual Property and Market Structure in Agriculture', in R. Evenson, W. Lesser, V. Santaniello and D. Zilberman (eds).

Reardon, T. and S. Vosti (1997), 'Poverty environment links in rural areas of developing countries', in S. Vosti and T. Reardon (eds.), *Sustainability, Growth and Poverty Alleviation*, Baltimore: IFPRI and Johns Hopkins University Press.

Robinson, W.I. (2000), Personal communication, Centre for Arid Zone Studies, UK.

Rukuni M., M. Blaickie and C. Eicher (1997), 'Crafting Smallholder-driven Agricultural Research Systems in Southern Africa', *Staff Paper No. 97–49*, Department of Agricultural Economics, Michigan State University.

Ruttan, V. (1997), 'Sustainable Growth in Agricultural Production: Poetry, Policy and Science', in S.A Vosti and T. Reardon (eds), *Sustainability, Growth and Poverty Alleviation*, Baltimore: IFPRI and Johns Hopkins University Press.

Santaniello V. (2000), 'Biotechnology and Traditional Breeding in Sub-Saharan Africa', paper presented at the workshop 'Biotechnology, Environmental Policy and Agriculture', European Science Foundation, UK Department of International Development and IPGRI, 29–30 May, Rome.

Traxler G., J. Falck-Zepeda and G. Sain (1999), 'Genes, Germplasm and Developing Countries' Access to Genetically Modified Crop Varieties', in R. Evenson, W. Lesser, V. Santaniello and D. Zilberman (eds).

World Bank (1991), *The Population, Agriculture and Environment Nexus in Sub-Saharan Africa*, Washington, DC: World Bank.

Zilberman D., C. Yarkin and A. Heiman (1999), 'Knowledge Management and the Economics of Agricultural Biotechnology', in R. Evenson, W. Lesser, V. Santaniello and D. Zilberman (eds).

3. The impacts of GURTs: agricultural R&D and appropriation mechanisms

Timothy Swanson and Timo Goeschl

INTRODUCTION

This chapter explores the nature of the changes that are occurring within the agricultural industry with the advent of so-called 'terminator technologies'. It argues that these technologies represent a systematic shift away from the current mixed (public/private) system for funding and directing research and development within agriculture, and towards a wholly private system for the same. Once these technologies are allowed to proceed, there will be no capacity to continue under the current mixed system. This change of regimes would confer some clear advantages. It would probably dramatically increase aggregate spending on R&D in agriculture, and it would probably increase it disproportionately in precisely those places that receive least expenditure under the current system. It would be anticipated that such increased levels of investment would also result in dramatic changes in agriculture, and the value of its production.

However, the loss of a public sector component to agricultural R&D is also an important institutional shift for very different reasons. First, it is likely that the private sector alone would not place as high priorities on certain issues crucial to agriculture, such as medium-to-long term sustainability or poor people's food security. Second, the shift to the private sector is motivated by the use of market-based mechanisms for determining the distribution of agricultural value. The current distribution of agricultural value is the result of complex and continuing negotiations. The shift to market mechanisms would discontinue these negotiations and displace this distribution.

This argument develops in the next three sections. The first of these discusses the R&D sector in modern agriculture, how it operates and the contributions of various important inputs (including genetic resources). The next discusses the property rights regimes applied to plant varieties, and how they determine both the incentives for investment and the distribution of value. The third section relates these concepts of industrial structure and incentive mechanisms to the current mixed system (public/private) and to the potential private one, developing the ideas discussed above. We conclude the chapter by outlining the range of possible impacts that biotechnologies might have on the industrial structure of agriculture.

INDUSTRIAL STRUCTURE: THE R&D PROCESS, ITS COMPONENTS AND THEIR CONTRIBUTIONS

R&D in Agriculture

Research and development (R&D) is the term used to describe the process by which new ideas are developed for application to common problems. When a new solution concept is successfully developed within the R&D process, it will then be marketed, usually embodied within some novel product. The industries that rely upon genetic resources are usually those that use R&D to solve certain recurring problems at the interface between human technology and the biological world.

Economists have long analysed the research and development process as one of information creation, application and diffusion (Arrow, 1962). The theoretical concept of the R&D process is usually represented as a production process itself dependent upon the application of various factors of production (machinery, labour and so on) for the production of useful ideas. Certain industries by their nature expend substantial proportions of their total available resources on the R&D process. These are those industries that have the creation of new information at the core of their functions. For example, the computer software, plant breeding and pharmaceutical industries are all R&D-intensive industries, with over 10 per cent of their gross revenues invested in the development of solution concepts. In a recent survey of the plant breeding industry, the surveyed breeders stated that they allocated, on average, 18 per cent of their annual turnover to research activities (Swanson and Luxmoore, 1998).

It is in this industry (and, to a related extent, the pharmaceutical industry) that the industrial R&D process most clearly depends upon the use of genetic resources in its search for solution concepts. Agriculture may be conceived of as a living defence system rather a static technology. In agriculture we continue to maintain a system that attempts to keep at bay the always evolving pests and predators of our primary food crops. The defences are neither absolute nor perpetual; they are constantly eroding under the pressure of the forces of natural selection. The R&D processes in agriculture are focused on providing solution concepts to the problems that arise in these contexts. The primary reason given for continued plant breeding is now the maintenance of resistance to pests rather than the enhancement of mean yields: 55 per cent versus 35 per cent of respondents in a recent survey reported this (ibid.).

Genetic resources are crucial inputs into this R&D process. The same forces that are at work against the human domain are also operating against all other life forms. Any organism that persists must do so because it has evolved 'successful strategies' in the sense that they are successful in a contested environment, that is, in resistance. Genetic resources are important inputs to R&D in these industries simply because they contain information which has been generated within the relevant crucible. It is not any genetic material per se that is the most useful input into these industries; rather it is the information to be gained from the characteristics which have evolved within a living environment that is most likely to make a contribution. Genetic resources have been useful in the past because of the manner in which the existing set of life forms have been selected (within a living, contested system similar to our own), which provides us with an already vetted library of successful strategies.

As the technological frontier expands, it becomes possible to make use of more and more discrete pieces of genetic material for their informational content, and to transport this information across greater biological distances. In the not too distant past, biological barriers prevented the transport of very precise biological functions across any distance, and no function was transportable across species. The advance of biotechnology has enabled the transport of very precise biological functions across vast biological distances. This technological advance enables the consideration of a much broader range of genetic materials as inputs into the solution of any given biological problem, but it does not alter the basic nature of the solution concept or R&D process.

Usually the information within nature requires substantial analysis and modification before it is incorporated within a final product. In this case the information from genetic resources is best considered as a raw informational input into the R&D process. That is, it is only after it is combined with other forms of capital (scientists, specialized equipment and so on) that the naturally generated information is able to be developed into useful applications. It is in this process that the intellectual input of various scientists becomes mingled with the information supplied by nature, and the final output is a composite of both.

Figure 3.1 illustrates the traditional approach to the generation of innovation in agriculture, commencing with the generation of new strategies in nature, their observation and selection by traditional farmers, and then their incorporation and amalgamation into widely used varieties by modern plant breeding companies. It is this flow of information that is accumulated and incorporated into the innovation we know as a modern plant variety.

Vertical Structure

In industrial economics, the term *vertical industry* is applied to the chain of production required to move the product from the stage of initial idea through production and into the hands of the consumer. Figure 3.1 sets out

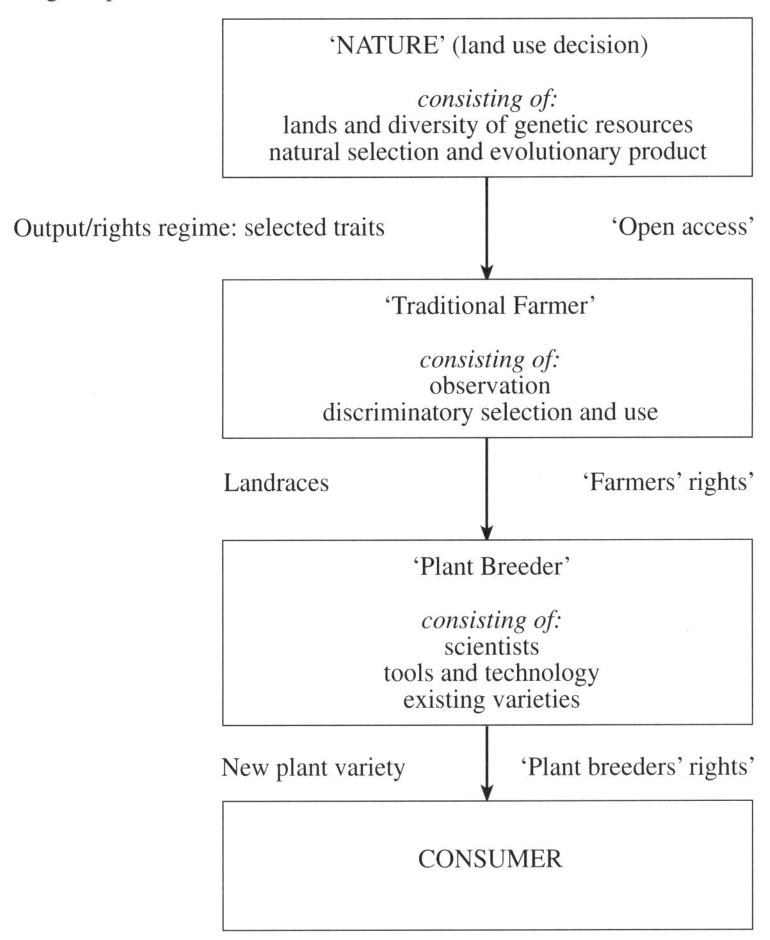

Figure 3.1 Vertical industry for plant breeding

a depiction of the vertical structure of the industry developing new plant varieties for agriculture (the 'plant breeding industry'). It depicts (starting from the top) how a flow of information originates within the natural environment by reason of the investment in certain inputs (lands and diversity) and that this flow of information is then channelled towards the market via the farmers (the intermediate stage of the industry) and finally to the plant breeders. These plant breeders then prepare and market the information to consumers. Each level of the industry must invest in certain forms of R&D activities in order to generate and appropriate the information flowing from the previous level; this consists of land use investments at the base of the industry and investments in scientists and laboratories at the market end of the industry. In this schematic there is no production at the end of this pipeline without flows occurring throughout its entire length.

In the first stage, effective characteristics for new plant varieties develop naturally through the process of 'natural selection': only those which are able to survive existing threats (pests and pathogens) remain and reproduce. Since the set of threats is constantly changing, the natural environment continuously produces new information on the characteristics that are relatively fit under current conditions. The maintenance of a relatively greater diversity of genetic resources and the dedication of greater amounts of lands to the retention of that diversity are the investment choices which determine the amount of information flowing out of this stage of the industry on the nature of the plants that work effectively in the prevailing environments.

The next stage of the industry consists of the individuals who observe the natural process of selection and aid in the dissemination of its information. 'Traditional farmers' have themselves survived by means of a process of observing this naturally produced information and the disproportionate use and transport of those plant characteristics which have aided survivability. They invest in the production of this information both by means of their land use decisions (as mentioned above) and by dedicating their time and resources to the observation and discriminatory use of those genetic resources which are revealed by nature to be of greater fitness. Their choices each year result in the capture of some of the flow of information on what was successful in the environment prevailing in the current year. This information also accumulates as a 'stock': traditional plant varieties (landraces) encapsulate the accumulated history of the information which nature has generated that farmers have observed and used disproportionately (Swanson and Goeschl, 1998).

At the end of this process, the 'plant breeding industry' has collected the set of varieties that farmers have created over millennia and hence the stock of naturally produced information that is encapsulated within them. By investing in laboratory equipment and scientists, the breeding process becomes focused on the use of this set of information for the preparation of the best possible variety for current environmental conditions. The modern plant breeder has then used its investments to create a variety that is an amalgam of some subset of the traditional varieties.[1]

To what extent is the stock of information within agriculture (represented by the accumulated set of landraces in storage) adequate to deal with the problems arising in agriculture at present and in the future? Although the past 30 years have seen an unprecedented rate of investment in the storage and general availability of landraces, it is interesting to note that the plant breeding industry has continued during this period to make use of 'new information', that is information that is coming in from the natural environment (Swanson, 1996). Hence the entirety of the vertical industry remains relevant to the maintenance of agriculture; it is not simply a matter of using the stock of information that we have accumulated but also a matter of managing the flow of information that is currently arriving.

Segregating between Contributions

This specific question may be addressed empirically by segregating between the returns attributable to the various factors of production in an industry such as plant breeding. A monopoly in marketing rights (IPR) provides the holder with the capability to receive a return on all of the new information embodied within the product, irrespective of its source. This means that an IPR system gives a rate of return on information that was produced by the rightholder's R&D process as well as on any other source providing information that is incorporated within the product. For example, in the context of the plant breeding industry (Figure 3.1), a plant variety certificate would provide a rate of return on all of the new information contained within a new plant variety, irrespective of whether it is generated by (a) the plant breeder's research and development process, (b) the traditional farmer's observation and selection process or (c) the natural environment's selection process.

The denomination of the property rights holder will determine the identity of the person receiving the return to the information contained within the product. For example, even if the plant breeder does nothing other than incorporate the information developed through other processes (for example, by crossing a landrace with a modern variety to incorporate its desired characteristics), the full value of this information will be captured by the holder of the property right. Here we assess the extent to which it is possible to ascertain the distinct contributions from these various factors of production within the plant breeding industry. This is done by specifying an 'R&D production function', and then estimating the extent to which its various component parts have contributed to the past production of new information. An R&D production function in the context of plant breeding, for example, would have to consist of at least (a) the scientific input (human capital), (b) the technological input (physical capital) and (c) the genetic resource input (natural capital). The theory of a production function states that increases in these various inputs would result in increases in the desired output: new modern plant varieties (Evenson and Gollin, 1991).

There has been at least one empirical study which has made an attempt to estimate the relative contribution of genetic resources in the R&D process in plant breeding. This study (Evenson, 1995) specified the R&D production function as follows:

$$\text{new varieties} = f(L, K, G)$$

where:
L is the level of input from human capital (scientists),
K is the level of input from physical capital (technology, machinery),
G is the level of input from genetic capital (biological diversity).

The empirical study was based upon the record of plant breeding at the International Rice Research Institute since 1960, and estimated the extent to which new varieties of rice were attributable to the various forms of investments. This study estimated that approximately 35 per cent of the production of modern new rice varieties has been attributable to the genetic resource input into the R&D function (ibid.). This implies that the inputs supplied by plant breeders in rice breeding (human and technological) generated no more than 65 per cent of the useful information within modern plant varieties. The imputed present value of a single landrace accession according to this study was $86–272 million. The imputed present value of 1000 accessions with no known history of use was $100–350 million. Given that the initial stock of rice germplasm (in 1960) was 20 000 accessions, the added stock of germplasm since that time (about three times as many accessions) has been estimated to be responsible for fully 20 per cent of the green revolution in rice production (Evenson, 1995).[2] Clearly, there are many important inputs to the generation of agricultural value: natural, human and scientific. One of the most difficult issues in global agriculture is the creation of an effective system for distributing this value in accord with efficiency and equity considerations.

INCENTIVE SYSTEMS: INTELLECTUAL PROPERTY RIGHTS, THE DISTRIBUTION OF RENTS AND THE INCENTIVES TO SUPPLY R&D

Intellectual Property Right Regimes as Incentive Mechanisms

The 'property right regime' that is often used when an industry is focused on the production of useful information is usually termed an *intellectual property right regime*. This regime is a social construct whereby returns are created to reward informational investments.

When R&D is a significant part of the production process within an industry, it is not always possible to obtain a reasonable rate of return on the product without an extended right of control over its subsequent use and marketing. This is because the end result of the R&D process is an idea, and this idea is then embodied in the products in which it is sold, and potentially lost on first sale. For example, a computer program that balances a bank statement is first an idea, and then a specific list of computer instructions created to effect that idea. If there is no exclusive right to control the subsequent marketing of the good (or close facsimiles thereof), the first purchaser of that good will have the right to produce competitive products without expending all of the R&D resources required to produce it initially. The first sale of the computer code in the previous example would enable the purchaser to make a similar program and set up in competition with the first. This is a problem if the first seller invested years in the construction of the program while the second only invested the few minutes (and dollars) required to copy it. In industries in which a substantial amount of the value produced is attributable to the information it contains (generated through R&D), there will be no incentive to invest in this R&D in the absence of the capacity to control the marketing of its goods even after their transfer to others. Intellectual property right regimes are analysed by economists as incentive mechanisms which give extended rights of control over the marketing of certain goods in order to provide incentives for the information-generating investments (R&D) that resulted in them (Arrow, 1962; Swanson, 1995b).

There is a great variety of rights denominated 'intellectual property': trademarks, copyrights, patents, plant variety rights and so on. The most important thing that all of these rights have in common is that they allow the holder to control some of the uses of the good subject to these rights *even after the good has left the rightholder's possession*. Thus a person with a copyright in a book is able to sell the book but retains the exclusive right to copy it; a person with a patent on a machine is able to sell the machine while retaining the exclusive right to manufacture it; a person with a registered plant variety certificate is able to sell that plant while retaining the exclusive right to reproduce it for resale.

The function of this extended right of control is to vest the holder with an exclusive marketing right in the particular good, usually for a limited period of years. This allows the holder to obtain a reasonable rate of return on the book, machine, plant variety or other good that is subject to the recognized right. Note that this rate of return is only available to the extent to which users recognize and enforce this right after the good has already left the possession of the rightholder. To the extent that the other users are willing to purchase from prior purchasers, the rightholder's exclusive marketing right will be of little value. There is a substantial increase in the rate of return afforded by allowing rightholders to control the uses of their rights outside of their possession. Rightholders term the unwillingness of other users to enforce their exclusive marketing rights 'piracy'.

Exclusive marketing rights are not required to earn a reasonable rate of return on the manufacture and sale of most goods. The vast majority of goods are sold without being subject to any such rights; the purchaser is usually within its rights to purchase the good and then commence producing similar goods after its purchase. For example, in the UK the patent-intensive industries produce only about 4.2 per cent of GDP and the copyright-intensive industries produce only about 3.7 per cent of GDP. Most goods and services do not require any extended right of control after sale; a reasonable rate of return is acquired, although all rights to use and production are transferred with the sale.

In addition, it must be noted that such marketing rights are second-best by definition. They are made to remedy one market imperfection (under-investment in information) by creating another (the creation of a limited term monopoly right). To this extent they will always be inferior to any system that is able to generate the information without the monopoly.

Property Regimes in Agricultural R&D and the Distribution of Rents

Plant varieties protection commenced in both the USA and Europe in the postwar era. It was harmonized and consolidated with the creation of the Union Internationale pour la Protection des Obtentions Végétales (UPOV), which was established in 1961 and came into force in 1968. This international convention established for the first time internationally recognized rights in registered plant varieties, so-called 'plant breeders' rights'. Patent rights in modified biological materials have also been allowed in recent years. The USA was the first jurisdiction to afford IPR protection to a living organism. The well-known case involved a bacterium useful in the petroleum industry that was the subject of the US patent ruling in 1980 first allowed the patenting of human-modified natural organisms (*Chakraborty* v. *Diamond*).

Figure 3.1 indicates at which levels of this industry property rights systems have (and have not) been developed. A large number of countries recognize the existence of plant breeders' rights. The registration of a plant variety by a seed producer under national plant registration legislation gives exclusive rights to the marketing of that variety, to the extent recognized by countries that are members of UPOV. The value of the information generated by the R&D process is appropriable by companies marketing a final product under this system of rights. At other levels of the industry there are ineffectual or non-existent rights regimes. The International Undertaking on Plant Genetic Resources recognized the concept of 'Farmers' rights', but they remain unimplemented. At the very base of the industry there has been until recently little or no recognition of the need to reward the flow of information that results simply from retaining a diversity of genetic resources on lands dedicated to that purpose.[3]

The existing rights structure will necessarily accumulate all of the returns from all of the investments in information at the end of the vertical industry. There are no other recognized rights to command a return, and so the only existing right is capable of capturing all of the returns from all of the earlier investments. For this reason there have been long and hard-argued international negotiations concerning the distribution of the returns from modern agriculture. The existing R&D system (public sector and private rights) is to some extent an always evolving, negotiated solution to this continuing dialogue.

Altered Property Right Regimes and the Incentives to Supply R&D

It would be expected that the impact of the introduction of a new property right system would be seen in (a) an increase in the investment in R&D in the affected industries, and (b) an increase in the investment in the inputs required to undertake R&D in these industries.

There is some evidence that the recent introduction of the new property right systems affecting genetic resources, and the expansion of others, have had some positive effect on the flow of information within these industries. The screening of microorganisms has been occurring at unprecedented rates in recent years, partly in response to the anticipated increase in appropriability of any information discovered in this process and partly in response to the reduced costliness of the screening process (Aylward, 1995). It is clear that the expansion of IPR systems into this area has increased private investment in R&D relating to these organisms, but further research is required to assess and analyse this link.

There is clear evidence that the introduction of IPR systems in the plant breeding industry (described above) expanded private sector R&D in this area. Studies in the USA indicate that R&D expenditures in just one field of research (small grains) increased in real terms by more than a factor of 100 between 1960 and 1991 (Pray and Knudsen, 1994). In the USA, passage of the implementing legislation for UPOV occurred in 1970, and Table 3.1 below demonstrates how R&D increased in anticipation of the receipt of the exclusive marketing rights acquirable since 1970. The number of R&D programmes in this area, the total number of plant breeders and the aggregate amount of R&D expenditure all rose dramatically from 1970.

Table 3.1 Private research and development activity (wheat case study)

	1960	1965	1970	1975	1979	1991
Number of R&D progs	1	3	6	9	9	11
Number of private plant breeders	3				23	25
R&D expenditure (1989 $1000)	31	1036	3102	6480	6937	4241
Number of PVPCs (5 yr period)			17	35	41	61

Note: PVPC = Plant Variety Protection Certificate.

Source: Pray and Knudsen (1994).

It is also clear that investment in some of the inputs required for plant breeding expanded with the introduction of plant breeders' rights. There was an expansion in the number of scientists trained and operating in the field and, as mentioned previously, a marked increase in the amount of genetic resource stocks in storage (Huffman and Evenson, 1993). In general, increased investment was directed at methods of preserving genetic resources that were of known usefulness. This evidence indicates the increased importance placed upon R&D in this field once IPR protection became available.

Investment was not increased for all of the R&D inputs potentially affected by this IPR regime. First, consider the extent to which the adoption of the IPR regime in plant breeding has resulted in such investments in the developing world. As of 1995, there were seven developing countries which had enacted IPR regimes regarding plant varieties: Argentina, Chile, Uruguay, Colombia, Mexico, Zimbabwe and Kenya (Jaffe and van Wijk, 1995). At least three of the Latin American countries have developed sectors making significant investments in the R&D necessary to secure registered plant varieties (see Table 3.2). At present, only two of these countries (Uruguay and Chile) have significant seed exports. Only one of these countries (Chile) had a trade surplus in seed and plant varieties (ibid.).

Table 3.2 PBR titles granted in three developing countries, 1968–94

	Domestic	Foreign
Argentina	416	206
Chile	141	90
Uruguay	16	9

Source: Jaffe and van Wijk (1995).

Note the nationality of the various concerns involved in plant breeding registrations in the most important countries in the field. In each of these countries the market in hybrid varieties is dominated by multinational firms, while the self-pollinating varieties are dominated by domestic operations. It is the former group that has experienced most of the growth in R&D expenditure in these countries (see Table 3.3). This demonstrates that the domestic plant breeding sector in these countries is not really competing with the multinationals but dealing in a completely distinct set of resources which are of much less interest to that industry. The agricultural research industry continues to be very concentrated in a small group of multinationals, and they in turn have been largely focused on the large-scale hybrid varieties (those that are protectable, not via law, but by biology).

Table 3.3 Average R&D expenditure of plant breeding companies in Argentina

Specialization of firm	1986	1992	Growth (%)
Hybrids	1286	1900	48
Self-pollinating	180	186	3
Diversified	370	851	130

Source: Jaffe and van Wijk (1995).

Most of the investment in plant breeding was occurring in those varieties where the returns from investments were appropriable. In developing countries most efforts by multinational companies occurred in the hybrid variety sector, where no legal protection was required. In the sectors where legal rather than biological protection was afforded (non-hybrids), most of the investment was focused on developed country varieties and developed country problems.

THE CURRENT MIXED SYSTEM OF AGRICULTURAL R&D

The public sector involvement in agricultural R&D is long-standing and intensive. For several decades plant breeding was the sole province of government and university (and sometimes botanical gardens) to collect resources and undertake plant breeding. In the 1970s, the international public sector launched another layer of collecting and plant breeding known as the Consultative Group of International Agricultural Research (CGIAR) system of agricultural research institutes. International donors provided further assistance for national agricultural research systems (NARs), the system of developing country research institutes that interfaced with the CGIAR system and their own farmer groups. Through this international, publicly funded, set of research stations, genetic resources were collected and disseminated, and then the products of research (utilizing these resources) were returned and redistributed.

The current system is a composite of public investment and private investment based on property rights. As mentioned above, the past two decades have witnessed many changes in the structure of the property right systems in this area. Hybrid varieties have been used to appropriate returns on investments since the 1960s. Biotechnology has been seen as one of the frontier technologies since the early 1980s, and movements to expand rights in this area culminated in the adoption of the Trade Related Intellectual Property Rights (TRIPs) Agreement in 1994. The current system is an amalgamation of public sector historic involvement and private enterprise initiatives. However, the precise mix of public and private investment has been changing over recent decades.

Funding support for public sector agricultural research suffered a significant setback in the 1980s in both developed and developing countries. Alston *et al.* (1998:51) observe:

> Governments everywhere are trimming their support for agricultural R&D, giving greater scrutiny to the support that they do provide, and reforming public agencies that fund, oversee and carry out research. This represents a break from previous patterns, which had consisted of expansion of public funds for agricultural R&D. Private sector spending on agricultural research has slowed along with the growth of public spending in recent years, but the balance continues to shift in favour of the private sector.

In a study of 22 OECD countries with particular emphasis on Australia, the Netherlands, New Zealand the UK and the USA (which account for over 40 per cent of public agricultural R&D by OECD countries), Alston *et al.* (1998) found that growth in public agricultural R&D spending had slowed down significantly in the 1970s and 1980s, as reflected in Table 3.4.

Table 3.4 Real agricultural R&D spending in OECD countries, 1971–93

	Millions of 1985 international dollars[b]				Annual growth rates (%)		
	1971	1981	1991	1993	1971–81	1981–93	1971–93
Australia	238.7	281.9	307.8	315.4	2.1	0.3	1.2
Netherlands	134.7	202.1	216.6	229.7	4.2	0.9	1.6
NZ	114.3	133.8	110.2	107.3	2.2	−2.2	0.2
UK	274.5	371.0	364.4	370.8	2.6	−0.2	1.2
USA	1235.6	1620.4	2023.4	2054.3	2.4	2.3	2.1
Subtotal (5)[a]	1997.8	2609.1	3022.4	3074.5	2.5	1.4	1.8
Other OECD (17)[a]	2300.2	3104.3	3919.1	4054.9	2.9	2.1	2.6
Total OECD (22)[a]	4298.1	5713.4	6941.4	7129.4	2.7	1.8	2.2

Notes:
[a]Figures in parentheses indicate the number of countries included in the respective totals.
[b]Research expenditures denominated in local currency are first deflated to 1985 prices using local implicit GDP deflators taken from World Bank tables (1995) and then converted to international dollars (where one international dollar is set equal to one US dollar) using purchasing power parities.

Source: Alston *et al.* (1998).

These figures can be taken to be representative of all developed countries because OECD countries account for almost 90 per cent of all developed country R&D. Alston *et al.* (1998) report that, in real terms, between 1945 and the mid-1970s, in most developed countries public expenditures on agricultural R&D grew more rapidly than in the rest of the post-World War II period. Then, in the mid-1970s, rates of growth in public R&D outlays slowed quite markedly and, in the 1980s, public agricultural R&D expenditures generally stagnated or declined. In the 1990s, however, public R&D expenditures recovered, or began to increase again, but at more modest rates of growth than in the 1960s or 1970s. The five countries of particular interest had markedly slower average annual growth rates in public agricultural R&D expenditures than the OECD as a whole, but the general pattern of growth in the five countries was similar to overall growth among the OECD countries.

The slowdown in the growth of agricultural R&D spending has been a feature of developing countries as well. Global agricultural R&D spending disaggregated by developed and developing countries is presented in Table 3.5. Note that in 1981 the developing countries' share of agricultural R&D expenditures approximately equalled that of the developed countries, while in 1991 it substantially exceeded that of the developed countries. Hence the current system for funding R&D in agriculture has managed to direct significant amounts of research investment towards developing countries. It is also evident from the expansion of production in developing countries (over the past three decades) that these countries have received some distribution of the value of global agriculture.

Table 3.5 Real agricultural R&D spending in developed and developing countries, 1971–91

	Expenditures (millions of 1985 international dollars)		
	1971	1981	1991
Developing countries (131)[a]	2984	5503	8009
Sub-Saharan Africa (44)[a]	699	927	968
China	457	939	1494
Asia and Pacific (excl. China) (28)[a]	861	1922	3502
Latin America and Caribbean (38)[a]	507	981	944
West Asia and North Africa (20)[a]	459	733	1100
Developed countries (22)[a]	4298	5713	6941
Total (153)[a]	7282	11217	14951
	Average annual growth rates (%)		
	1971–81	1981–91	1971–91
Developing countries	6.4	3.9	5.1
Sub-Saharan Africa	2.5	0.8	1.6
China	7.7	4.7	6.3
Asia and Pacific (excl. China)	8.7	6.2	7.3
Latin America and Caribbean	7.0	−0.5	2.7
West Asia and North Africa	4.3	4.1	4.8
Developed countries	2.7	1.7	2.3
Total	4.3	2.9	3.6

Note:
[a]Figures in parentheses indicate the number of countries included in the respective totals.

Source: Alston *et al.* (1998).

It is also important to note that public sector involvement in R&D can rarely be justified on grounds of efficiency. Governments are notoriously bad in their judgment of the appropriate directions for innovation. However, in the case of certain industries (food and drugs being two of the best examples), governments are often responsible for the management of the rate and direction of change – in order to safeguard the health of the public. Under the current system, it might be argued that the pace and direction of innovation in agriculture are less important than the sustainability of innovation and the equitable distribution of its proceeds.

The distribution of the benefits from agricultural R&D is also an issue that is publicly negotiated. At present, the 1991 UPOV agreement provides for a farmers' exemption for storing seed and replanting. This mirrors the 'own use' proviso within copyright law, and allows each farmer to make his own decision regarding the trade-off between food security and productivity. The CGIAR system also fosters this approach by means of disseminating freely its plant varieties, and by working with and contributing to the system of NARs. All of these public sector investments are ways in which global product is redistributed towards the less technologically capable countries.

THE POTENTIAL IMPACTS OF GURTs IN AGRICULTURAL R&D

The agricultural R&D industry underwent a significant change with the advent of hybrid variety-based lines. Since that time the industry has been able to appreciate the full extent of the benefits it might be able to reap from the fruits of its enterprise. All that was required was a mechanism that enabled appropriability in other varieties.

The first efforts by the industry were focused on legal changes, but investment patterns indicate that the appropriability of benefits from plant breeders' rights never matched that attained in the hybrid varieties. Another feature of the industry probably explicable by such incentives is its level of concentration. The high level of concentration is probably another attempt to capture or minimize externally conferred benefits. However, the concentration of the plant breeding/seed industry has not been able to enhance appropriation in the absence of new technologies, since farmers alone could replicate the product. The pursuit of so-called 'terminator technologies' is an attempt by the industry to resolve these problems of appropriability, in order to justify the massive investments in biotechnologies that are set to occur. They are the technological outcome of several decades of efforts by the industry to resolve this problem of appropriability.

The Impact on the Agricultural R&D System

What would be the logical outcome of this change in incentive systems? It seems possible that there would be several linked outcomes. We postulate the following impacts, and indicate why these might be anticipated.

1. *A dramatic increase in private R&D expenditures on agriculture* The increased appropriability of the returns from agricultural improvements in general would increase incentives to invest in R&D. Since IPR systems are necessarily imperfect means of appropriating returns, a technology-based system would enhance appropriability and hence increase the incentives to invest in agricultural R&D.
2. *Disproportionately greater increase in appropriability of returns within developing countries* The current (IPR-based) incentive system provides for very low rates of appropriability in developing countries, because those countries have the capacity to determine the degree of enforcement of the system within their borders. In many developing countries plant breeders' rights and patent systems are seldom enforced. This is because these countries do not see any direct benefits to be received by enforcing these rights against their own farmers, owing to the absence of an indigenous plant breeding sector. (In addition, plant breeders' rights systems do not require enforcement against farmers' use of plant varieties, so long as the seed is not marketed.) In those countries where little appropriability has been possible in the past, GURTs would represent a qualitative leap in appropriation.
3. *An incentive to increase private R&D expenditures on varieties suitable for use in developing countries* Any system that enhances the appropriability of returns within developing countries – without the necessity of having recourse to the legal system – will have a disproportionate impact on appropriation from varieties for developing countries. This would provide the private sector with the incentive to redirect some of its R&D towards the needs and objectives of countries in the developing world.
4. *A much narrower and more uniform R&D sector in agriculture* The R&D sector in agriculture has historically been a very diverse system (farmers, state NARCs, international agencies) and the flow of germplasm for research has been relatively free and unrestricted. As a side-effect, GURTs is likely to remove the flow of germplasm from the public arena, and make it a matter of private commerce. This will result in a single private property rights-based system of R&D. This is because the new system would create enforceable property rights that

must be acquired prior involvement in agricultural R&D. Once private licences are required, the costs of public sector involvement in agricultural R&D will escalate. With the passage of time, there will be little or no capacity to undertake R&D in the industry without the initial acquisition of a large portfolio of licences. This would create a significant barrier to entry for the more informal participants (farmers, NARCs) in the current system. Equally importantly, the benefits from any varieties that the public sector would create would then be similarly appropriable (via the new technologies) and it would make little sense for the public sector to release these freely while needing to pay for licences for others. With the passage of time (after GURTs), it is highly likely that the two sectors (public and private) would converge (on a licence-based system) and the informal parts of the R&D system would be removed.

The Impact of GURTs on Developing Countries as a Group

The impact of this change of systems on developing countries is not clear-cut, nor is it likely to be the same for each and every developing country. However, there are a few general impacts that can be noted.

1. *The loss of a significant portion of the existing diverse R&D industry and its outputs* The expansion of the private R&D sector would come at the expense of the loss of the existing diverse and informal one, especially farmers and traditional plant breeders. This would result in the loss of an existing flow of information from the informal sector to the formal one, and it might result in the loss of a flow of innovations from the private sector through the informal sector to farmers on the fringe of the commercial sector.
2. *A loss of public control over the direction and diffusion of agricultural R&D* The loss of public sector involvement would eliminate the public capacity to direct agricultural R&D toward the crops and pests indigenous to developing countries. Agricultural R&D would then be directed toward those crops and pests dictated by commercial concerns. There may be private-based incentives to broaden R&D objectives to include developing country problems, but past experience does not indicate that this is necessarily the case.
3. *The loss of control over the negotiated distribution of the value of agricultural R&D* A complex and continuing process of international negotiations has determined the current distribution of the value of agricultural R&D. The pattern of public sector spending and private sector rights in agricultural development represents one of the ultimate

outcomes of these negotiations. As public sector spending in agricultural R&D is 'crowded out' by the expansion of the private sector, there will no longer be any other obvious point for international intervention. Then the private market will fully determine the distribution of rents. In effect, there is the possibility that all existing systems/agreements for distributing these rents will be abrogated and replaced by private property rights-based distributions.

4. *Increasing concentration and control of agricultural R&D: potential market power* The corollary of the loss of diverse involvement in agricultural R&D is its increasing concentration in a small number of large biotechnology firms. This concentration could result in the increasing capture of rents resulting from market power as well as from innovation. Rent capture due to market power is pursued by means of the restriction of output, and this implies reduced innovation and reduced diffusion of innovations.
5. *A loss of public control over the objectives of agricultural R&D* The loss of public sector involvement would eliminate the public capacity to determine the fundamental objectives of agricultural R&D. The future direction of agricultural research would be determined solely by profit-based incentives and market opportunities. There is good reason to be concerned about exclusive reliance on market-based incentives for giving priority to sustainability and food security as agricultural issues. Discount rate arguments alone indicate that there is good reason for observing caution before relegating these problems to the private sector.

Determining the Net Impact of GURTs

The net impact of GURTs will be the outcome of these two distinct impacts: the *potentially positive impacts* derived from enhanced appropriability versus the *potentially negative impacts* derived from dramatic changes to the industry's structure. As set out above, the benefits of enhanced appropriability should generally result in increased levels of innovation in agriculture, and hence increased levels of productivity. In the first instance, this enhanced appropriability should be greatest in those countries with the lowest levels of IPR enforcement, and hence levels of innovation should be increasing in the developing as well as the developed world.

As set out above, the industrial structure in agricultural R&D is likely to change dramatically, from the existing diverse and diffuse structure to one that is homogeneous and concentrated. This new structure would have several potentially detrimental effects: the loss of public sector involvement, the loss of diverse flows of innovation, the loss of ancillary methods of diffusion and, potentially, the restriction of the level of innovation and diffusion.

The Impact of GURTs on Individual Developing Countries

The impact on individual developing countries will be dependent on the extent to which the individual country is able to direct and use biotechnologies (BT). If a country is fully BT-capable, it is likely to acquire some of the rents from this new approach to R&D, and it is likely to 'catch up' with other countries' innovations relatively rapidly. If the country is wholly BT-incapable, it is unable to capture rents within the industry and it is unlikely to be able to 'catch up' quickly with other countries' innovations. The impact of GURTs is likely to vary across this entire spectrum.

Developing countries with BT capability should benefit unambiguously from the introduction of GURTs. They will be able to protect their innovations and capture enhanced benefits from them. In addition, countries with BT capability will be much better equipped to 'catch up' with innovations made by others. They will need BT capability in order to have any prospect of 'reverse engineering' GURT-protected varieties, and then translocating their innovations into other plant materials. Although the extent of such reverse engineering is debatable, it is likely that the rate at which the diffusion of innovations (to a given country) occurs will depend in part on its capability in biotechnology.

The impact of GURTs is ambiguous for those countries without BT capability. In the past it was always possible to 'catch up' with others' innovations through traditional plant breeding activities. Since the plants with innovative characteristics continued to reproduce, it was possible to interbreed them with other plant materials. It was also possible to make use of the resulting new varieties without being in violation of existing plant breeder rights. GURTs will render these forms of plant breeding activities infeasible, and hence eliminate these *direct* forms of diffusion. These countries will therefore receive the impact of GURTs only *indirectly*, that is, only to the extent that the BT-capable firms alter their rates of innovation and diffuse those innovations to the BT-incapable.

This indicates that one important parameter in determining the extent to which a country will benefit from GURTs is the length of time until it becomes fully BT-capable. If every country was equally biotechnologically capable, many of the potential problems of GURTs would not exist. To the extent that a country is able to join the group of BT-capable, it is likely to remove many of the potential drawbacks of the technology, from its own perspective. (This is not intended to suggest that there are not other drawbacks that might continue from the social perspective, that might derive from the divergence of private and public objectives in agriculture.)

The Impact of GURTs within Individual Developing Countries

The impact of GURTs will vary not only between developing countries but also within developing countries. This is because the diffusion of innovation within a given developing country is not a uniform occurrence. There is pronounced 'dualism' within many parts of the developing world, and this means that the existence of an advanced technology sector does not necessarily redound to the benefit of all citizens uniformly. It is often the case that the benefits of high technology within a developing country do not diffuse any more quickly between these sectors within the country than they do between sectors outside the country.

BT-capable countries may have the capability to generate benefits from GURT technologies, and also to catch up with those innovations realized elsewhere; however, this does not necessarily imply that these benefits will diffuse equally throughout this country. This will depend on how the benefits of agriculture and agricultural innovation have been distributed in the past. In those countries where benefits have diffused very unequally in the past, it is unlikely that GURTs will generate a better distribution of benefits. Many countries have a 'dual agricultural system': a modern sector and a traditional one. To the extent that the innovations within GURT varieties are directed to the modern sectors, inequalities within countries will be as great as they are between countries.

NOTES

1. There are exogenous factors inherent in the nature of plant breeding and pharmaceutical development that may contribute to increasing the potential returns on investment in R&D when genetic information is involved. In particular, there may be cumulative effects in R&D such that the marginal productivity of R&D increases over time. The reason is that there may be (a) increasing returns associated with combining informational inputs and physical and human capital and (b) increasing returns associated with combining different informational inputs. The latter aspect is explored in depth in Weitzman (1998).
2. The studies conducted by Evenson and others used as a measure of 'genetic resource inputs' the number of plant varieties held within a public gene bank. Of course it is crucial that, for additional varieties to provide additional value, the varieties be dissimilar from those already held and be inclusive of proven resistance strategies (evolved in a natural system).
3. In this regard, the International Undertaking on Plant Genetic Resources propounded by the FAO Commission on Plant Genetic Resources has attempted to stake a claim to a return for the traditional farmer, but this remains to be implemented. In June 1998, the World Intellectual Property Organisation (WIPO) declared that it plans to extend the IPR

system to (what is termed) 'new beneficiaries'; that is, to create protection of rights of holders of traditional knowledge, indigenous peoples and local communities. WIPO remains unclear, however, about the nature of these rights and the mechanism through which claims would be assessed and settled (WIPO, 1998). Many of the benefit-sharing mechanisms being discussed elsewhere also concern this level of the industry, and the need to share benefits with those individuals who have created the stock of information on which we rely.

REFERENCES

Alston, J.M., P.G. Pardey and Vincent H. Smith (1998), 'Financing Agricultural R&D in Rich Countries: What's Happening and Why?', *The Australian Journal of Agricultural and Resource Economics*, 42(1), 51–82.

Aylward, B. (1995), 'The role of plant screening and plant supply in biodiversity conservation, drug development and health care', in T. Swanson (ed.), *Intellectual Property Rights and Biodiversity Conservation*, Cambridge: Cambridge University Press.

Arrow, K. (1962), 'Economic Welfare and the Allocation of Resources for Invention', in R. Nelson (ed.), *The Rate and Direction of Inventive Activity*, Cambridge, MA: Harvard University Press.

Calabresi, G. and D. Melamed (1972), 'Property Rules, Liability Rules and Inalienability', *Harvard Law Review*, 85, 1089–1128.

Coase, R. (1960), 'The Problem of Social Costs', *Journal of Law and Economics*, 1 (1), 15–45.

Evenson, R. (1995), 'The Valuation of Crop Genetic Resource Preservation, Conservation and Use', paper prepared for the Commission on Plant Genetic Resources, Rome.

Evenson, R. and D. Gollin (1991), *Priority Setting for Genetic Improvement Research*, Los Banos, Laguna, Phillipines: International Rice Research Institute.

Gollin, D. and R. Evenson (1990), 'Genetic Resources and Rice Varietal Improvement in India', Economic Growth Center, Yale University.

Grossman, S. and O. Hart (1986), 'The Costs and Benefits of Ownership: A Theory of Vertical and Lateral Integration', *Journal of Political Economy*, 94(4), 691–719.

Hart, O. and J. Moore (1990), 'Property Rights and the Nature of the Firm', *Journal of Political Economy*, 98, 1119–58.

Huffman, W. and R. Evenson (1993), *Science for Agriculture*, Ames: Iowa State University Press.

Jaffe, W. and J. van Wijk (1995), 'The Impact of Plant Breeders' Rights in Developing Countries', technical paper of the Special Programme on Biotechnology and Development Cooperation, Ministry of Foreign Affairs of the Netherlands, The Hague.

Juma, C. (1988), *The Gene Hunters*, London: Zed Books.

Kloppenburg, J. (1988), *First the Seed*, Cambridge: Cambridge University Press.

Michelman, F. (1971), 'Property, Utility and Fairness', *Harvard Law Review*, 80, 1165–1258.

Perrin, R., K. Kunnings and L. Ihnen (1983), 'Some Effects of the US Plant Variety Protection Act of 1970', Economic Research Report no. 46, Department of Economics, North Carolina State University.

Posey, D. and G. Dutfield (1996), 'Beyond Intellectual Property', International Development Research Centre, Ottawa.
Pray, C. and W. Knudsen (1994), 'Impact of Intellectual Property Rights on Genetic Diversity: The Case of Wheat', *Contemporary Economic Policy*, XII, 102.
Repetto, R. and M. Gillis (1988), *Public Policies and the Misuse of Forest Resources*, Cambridge: Cambridge University Press.
Sedjo, R. (1992), 'Property Rights, Genetic Resources and Biotechnological Change', *Journal of Law and Economics*, 35, 199–213.
Sedjo, R. and D. Simpson (1995), 'Property rights, externalities and biodiversity', in T. Swanson (ed.), *Intellectual Property Rights and Biodiversity Conservation*, Cambridge: Cambridge University Press.
Sedjo, R., D. Simpson and J. Reid (1995), 'Valuing Biodiversity for Use in Pharmaceutical Research', Resources for the Future Working Paper.
Shiva, V. (1991), *The Violence of the Green Revolution: Third World Agriculture, Ecology and Politics*, London: Zed Books.
Simpson, D. and R. Sedjo (1998), 'The Value of Genetic Resources for Use in Agricultural Improvement', in R.E. Evenson, D. Gollin and V. Santaniello (eds), *Agricultural Values of Genetic Resources*, Rome: FAO.
Smale, M. (1997), 'The Green Revolution and Wheat Genetic Diversity: Some Unfounded Assumptions', *World Development*, 25(8), 1257–69.
Swanson, T. (1995a), 'Uniformity in development and the decline of biodiversity', in T. Swanson (ed.), *The Economics and Ecology of Biodiversity Decline: The Forces Driving Global Change*, Cambridge: Cambridge University Press.
Swanson, T. (1995b), *Intellectual Property Rights and Biodiversity Conservation*, Cambridge: Cambridge University Press.
Swanson, T. (1996), 'Biodiversity as Information', *Ecological Economics*, 17, 1–8.
Swanson, T. (1999), *Biodiversity Conservation via Alternative Pathways to Development. Biodiversity and Conservation*, 15, 83–115.
Swanson, T. and T. Goeschl (1998), 'Optimal Management of Genetic Resources for Agriculture: Ex Situ and In Situ', *Journal of Agricultural Economics,* special issue – papers and proceedings of World Congress.
Swanson, T. and A. Kontoleon (2000), 'Nuisance', in B. Bouckaert and G. De Geest (eds), *Encyclopedia of Law and Economics, Volume II*, Cheltenham, UK and Northampton, USA: Edward Elgar, pp. 380–402.
Swanson, T. and R. Luxmoore, R. (1998), *Industrial Reliance Upon Biodiversity*, Cambridge: WCMC.
WCMC (1992), *Global Biodiversity*. London: Chapman and Hall.
Weitzman, M. 1998, 'Recombinant Growth', *Quarterly Journal of Economics*, 113(2), May, 331–60.
World Intellectual Property Organisation (WIPO) (1998), 'Documentation of the WIPO Roundtable on Intellectual Property and Indigenous Peoples', 23 and 24 June, WIPO/INDIP/RT/98/1-4, Geneva.

4. Agricultural biotechnology and developing countries: proprietary knowledge and diffusion of benefits

Charles Spillane

INTRODUCTION: THE DISTRIBUTION OF THE BENEFITS OF AGRICULTURAL BIOTECHNOLOGY

The development of a technology (especially of a commercial and proprietary nature) does not imply the widespread dissemination of the technology, especially to poorer social groups who do not represent an immediate commercial market where R&D costs can be recouped (Byerlee and Fischer, 2000). This is one reason why it is as yet more difficult to identify many instances where modern biotechnological research has been practically applied in the fields of poorer farmers in developing countries than it is to simply list the range of possible improvements (Royal Society, 2000).

There are competing claims regarding agricultural biotechnology and its potential contributions to food security (most discussion is limited to the context of Malthusian projections regarding population demand and food supply). In July 2000, academies of sciences in Brazil, China, India, Mexico and the United States, the Third World Academy of Sciences in Trieste and the Royal Society met to debate the new. Their report calls on private companies that have developed GM technology to use their expertise to help the poor, and on governments to maintain publicly funded research in the field. The academies do not intend to indicate that agricultural biotechnology is the only solution to world hunger, but assert that it can make a significant contribution. The document was a consensus of opinions from the Royal Society, the US National Academy of Sciences, the Third World Academy of Sciences and the science academies of China, Brazil, India and Mexico (*www.royalsoc.ac.uk*).

There are already quite a few examples of the positive impact of plant tissue culture and micropropagation on the livelihoods of resource-poor farmers (Qaim, 1999a, 1999b; Thro and Spillane, 2000), yet the unfortunate reality is that the agricultural biotechnology research community has to date few examples where pro-poor applications of modern molecular biotechnologies are (or are likely soon to be) in evidence in farmers' fields on a significant enough scale to have a major impact on rural poverty (Royal Society, 2000). Amongst others, the following are some of the better known examples that are currently in (or close to application in) poorer farmers' fields in developing countries:

- transgenic rice with higher levels of vitamin A or iron (Ye *et al.*, 2000; Goto *et al.*, 1999);
- transgenic rice resistant to rice yellow mottle virus in West Africa (Pinto *et al.*, 1999);
- striga resistant transgenic maize varieties in Kenya (Gressel *et al.*, 1996);
- papaya resistant to papaya ringspot virus in Hawaii (Gonsalves, 1998);
- about 0.3 million hectares (1 per cent of global area) of transgenic crops in China (James, 1999; Pray *et al.*, 2000);
- virus-resistant transgenic sweet potato in Kenya (Qaim, 1999b).

While the paucity of examples in poorer farmers' fields at this stage may reflect the novelty of the more advanced technologies or the increasing regulatory barriers to their field-level implementation, it may also simply reflect a commercial research bias towards the demand of richer farmers and consumers. This chapter surveys some issues regarding the proprietary nature of biotechnological advances and explores some questions regarding the impact of these property rights on the rates of diffusion of the benefits to the poorer farmers worldwide.

Over the past decade there has been some advocacy of a need for a pro-poor bias in the development and dissemination of modern biotechnologies (Bunders, 1990; Swaminathan, 1991; Lipton, 1999). However, whether any benefits of current agricultural biotechnology research will actually reach poorer farmers and consumers in the near term without major public sector intervention is an open question (Spillane, 1999, 2000). There are many such poorer people. Over 1100 million farmers, in many different farming systems and environments, are economically active in agricultural production globally (about 50 million farmers in the developed countries and 1050 million in the developing countries).

The vast majority of the world's farmers are known to have limited access to external inputs or other productive resources. Resource-poor farmers, by definition, are unlikely to have easy financial access to agricultural inputs such as pesticides, fertilizers or irrigation. Moreover, it is now thought that an increasing majority of the world's resource-poor farmers are women. For instance, over 70 per cent of the people in developing countries living below the poverty line are women, the majority of whom live in rural areas (UNDP, 1999). Food staples typically absorb half the consumption of people below the poverty line (Lipton, 1999). The low productivity of resource-poor farmers tends to perpetuate rural poverty to the extent that, of the more than 2500 billion people in developing countries who live in rural areas, approximately 1000 million live below the poverty line: 633 million in Asia, 204 million in Africa, 27 million in the Near East and North Africa, and 76 million in Latin America (Jazairy *et al.*, 1992).

While such resource-poor farmers practise approximately 60 per cent of global agriculture, they produce 15–20 per cent of the world's food (Francis, 1986). However, when looked at another way the small-scale resource-poor farming sector is responsible for 80 per cent of agricultural production in developing countries and is therefore the key to future food security (Daw, 1989). Although the percentage of economically active people engaged in agriculture can vary widely between countries (for example, 92 per cent in Nepal, 2 per cent in USA), on average almost 50 per cent of the world's economically active people are engaged in agriculture. In many developing countries, poorer consumers and poorer farmers are one and the same.

Science alone is unlikely to provide a 'technical fix' for alleviating rural (or urban) poverty. There are many processes, factors and socioeconomic structures underlying rural people's poverty, such as lack of access to land and other productive resources, low purchasing power, political powerlessness, fragile environments and peripherality from markets (Sen, 1981). In this milieu, agricultural (or indeed biotechnology) research is but one factor among many which could have differential impacts on rural poverty. Indeed, the potential contribution of biotechnology to developing country agriculture or to poverty alleviation is considered to have been overstated, in the short term at least (Brenner, 1996). Yet over the longer term there is little doubt that some biotechnological approaches to agricultural improvement could generate social, economic and environmental benefits if aimed at specific needs, especially those of poorer groups.

Such needs might, for instance, include staple crops with higher yield potential; reduction in pesticide use via insect/disease-resistant crops and animals, veterinary vaccines (for example against brucellosis, rabies, encephalitis, liverfluke, hepatitis), improved nutritional composition of crops and animals, elimination of toxic substances or allergens, developing early maturing varieties, reducing post-harvest storage losses, abiotic stress-tolerant crop and animals, varieties and breeds with increased water and nutrient use efficiency, enhanced nutritional and product quality through influencing quantity and quality of oil, protein, carbohydrates and nutrients, and novel substances. While a vast range of approaches for the biotechnological improvement of product-oriented agronomic traits are either under study or in early development phases, given the current lack of focused public sector support for pro-poor agricultural biotechnology it is unlikely that poorer farmers will have economic access to such traits over the near term.

Although they are interrelated, some of the most neglected needs of the poor are of a 'process' rather than 'product' nature, for instance reducing labour demands (or increasing labour productivity) at appropriate times during the cropping cycle of smallholders whose main resource is their labour (Salamini, 1999). Technology which can deliver increases in land and labour productivity for resource-poor farmers are likely to be more adoptable. Resource-poor farmers constantly face difficult choices in their time–labour allocation decisions. However, for landless farmers their labour is often their main resource. Development of technology that relieves farmers' time burdens in agricultural production (and household maintenance) without sacrificing their ability to earn independent incomes is therefore critical (Carroll, 1992). It is worth considering the major role of children's work within agricultural households in this regard also. There are between 200 and 400 million child labourers throughout the world (UNICEF, 1997). The vast majority of children in resource-poor rural households are involved in significant levels of agricultural work (White, 1996). One of the greatest factors perpetuating poverty and population growth is lack of access to education. Labour demands on poorer children can limit their access to schooling.

For obvious commercial reasons, richer farmers are likely to be the main target market for most privately funded agricultural biotechnology research. Despite some rhetoric to the contrary, the many resource-poor farmers in developing countries who depend on an income of less than a dollar a day are not likely to be a near-term target market for most of the agricultural biotechnology companies. In addition, because over half of the people in the developing countries in 2010 will still look to farming for their livelihoods, a major challenge facing publicly funded agricultural biotechnology research which hopes to have a rural poverty alleviation objective will be to improve the labour productivity of agriculture in a manner which does not lead to job displacement or loss of income among poorer rural groups. In essence, if agricultural biotechnology is to help address problems of rural poverty and malnutrition it will have to purposely shift its focus from crops that feed chickens to the staple crops that feed poorer people; and from meeting the needs of large, low-employment farms to meeting the needs of smallholders and farm labourers (Lipton, 1999).

The major agricultural biotechnology companies are concentrating on the development of two broad types of proprietary traits: (a) input traits such as herbicide tolerance, insect or disease resistance and (b) output traits which improve the nutritional contents of foods or exhibit unique properties for very specific end uses or markets (Hayenga, 1998). Such input and output traits will be incorporated into existing elite varieties to provide seed with further added value, which may offer to the farmer lower costs or higher yield, and increased value of the end product. Initially, it is likely that biotechnology-generated varieties brought to the market will focus on input traits. However, the long-term commercial potential of plant biotechnology is considered to be in the development of value-added output traits that will address a wide range of specific needs or market niches (Shimoda, 1998).

To assess the level of demand for particular types of traits or products, the marketing departments of most agricultural biotechnology companies typically conduct surveys of farmers (or consumers) of different income levels to determine what commercial products might be developed by their technologists to meet the needs of those customers who can express their demand in financial terms. Because demand-driven companies are likely to make more money, most marketing departments in companies are highly responsive to customer concerns and demands. Within many public sector research organizations there is often an absence of demand-driven biotechnology research agendas, especially in relation to the agronomic or socio-economic needs of poorer farmers or developing countries (Ashby and Sperling, 1995). Some publicly funded plant biotechnology research could in effect be competing with the private sector for the same customers or clients, or working in a surrogate subsidized 'support' capacity for the private sector.

For the private sector, poorer farmers and consumers are by definition not a lucrative market and are unlikely to exert any effective 'demand pull' on the private sector research agenda. Those market niches or needs which are the most financially lucrative will exert the greatest demand pull on privately funded agricultural biotechnology research. In theory, the onus would therefore fall on the public sector to fund and perform any research required to meet the differential needs of poorer farmers or consumers. Public funding is usually scarce, and the competition between multiple objectives is intense. Since the supply of most agricultural biotechnology products is currently biased towards a commercial sector with narrowly defined financial objectives, there is little incentive to develop a broader portfolio of agricultural biotechnology techniques and materials.

This chapter proceeds as follows. The next section outlines the scales of private and public sector investments in biotechnology, and indicates the direction that these investments usually take. The third section surveys the general legal system that provides the system of proprietary rights, and the resulting system of incentives. The fourth section outlines how this system of proprietary rights affects the diffusion of biotechnologies within agriculture. The fifth section provides suggestions on how the diffusion of biotechnology's benefits might be enhanced through public intervention, while the sixth section concludes.

BIOTECHNOLOGY RESEARCH AND DEVELOPMENT INCENTIVES AND INVESTMENTS

There is little doubt that, if plant biotechnology research was applied to well defined social or economic objectives, it could benefit poorer farmers. Different groups of farmers will have different needs regarding agricultural biotechnologies (Leisinger, 1999) and hence any meaningful priority setting regarding research objectives is likely to be specific to particular countries, crops or groups of farmers (Brenner, 1996). However, there remains the valid concern that the needs of poorer farmers or nations are unlikely to be a factor which favourably steers the research objectives of biotechnology research which is wholly dependent on private investment (Barton and Strauss, 2000). Long-term public sector investment in agricultural (biotechnology) research will be essential to address the needs of poorer farmers and consumers who do not constitute a significant enough commercial market for private sector biotechnology research and development (Brenner, 1996).

Private and Public Sector Investment in Agricultural Biotechnology

The global market for agricultural biotechnology products was less than US$500 million in 1996, but is projected to increase to US$20 billion by 2010 (James, 1997). The world market for crop seed is valued at approximately $45 billion, which can be roughly divided into three main categories of equal size: commercial seed, farm-saved seed and seed provided from government institutions. Of the $15 billion global market in commercial seed at present, hybrids account for approximately 40 per cent of sales and reportedly most of the profit margins (Rabobank, 1994). The value of the global transgenic crop market was projected at about $2 billion for the year 2000, increasing to $6 billion in 2005 (James, 1997). Global sales from transgenic crops were estimated at $75 million in 1995; sales tripled in 1996 and again in 1997, to reach $235 million and $670 million. respectively, more than doubled in 1998 to reach $1.6 billion and increased by more than a third in 1999 to reach an estimated $2.1 to $2.3 billion (James, 1999). The global market for transgenic crops is now projected to reach approximately

$8 billion in 2005 and $25 billion in 2010 (ibid.). It is also projected that biotechnology-based solutions to weed, fungal and insect problems will soon comprise 10–20 per cent of the global $45 billion crop protection market (Hayenga, 1998).

The past decade has seen a major increase in private sector investment in agricultural biotechnology. Private sector agricultural research in OECD countries is now in excess of $7 billion and accounts for half the world's entire agricultural research investment. An increasing proportion of this agricultural research investment is in modern biotechnologies – an estimated $1.5 billion in 1996 (Byerlee and Fischer, 2000). In some of the countries where such private investments have been highest there has also been significant public sector investment in agricultural biotechnology research. At the international level, a number of public sector institutions are now assigning a higher priority to agricultural biotechnology. The World Bank has lent $100 million in support of biotechnology initiatives, whilst the Rockefeller Foundation and various bilateral donor agencies (for example, in the USA, the UK, the Netherlands) have invested $200 million in agricultural biotechnology over the past decade (Brenner, 1996). It is estimated that donors provide US$40–50 million per year for agricultural biotechnology initiatives in developing countries (Byerlee and Fischer, 2000). Such donor funding accounts for a high percentage of the total agricultural biotechnology research expenditures in most developing countries (excluding India, China, Brazil) for example, around 60 per cent for Kenya and Zimbabwe between 1985 and 1997 (ibid.).

The CGIAR (Consultative Group on International Agricultural Research) estimate that their biotechnology expenditures are approximately $22 million per year (about 7 per cent of overall CGIAR budget), of which only about $10 million is spent on crop biotechnology spread across eight different international agricultural research centres (IARCs). Considering the number of mandate crops studied by the CGIAR and their global importance in social terms, the CGIAR's expenditure on crop biotechnology is extremely small relative to some other large public or private sector agricultural research organizations. For instance, the USDA's 1995 expenditure on agricultural biotechnology research was $2 billion. In 1998, the CGIAR's Third External Review Panel concluded that there was a need for the CGIAR to better harness for the public good the advances taking place in agricultural biotechnology, in particular to ensure that the needs of the poor in developing countries are met (see *http://cgreview.worldbank.org/*). Whether such harnessing will be possible without increased public sector funding for CGIAR research on a crop-by-crop basis is an open question.

Biotechnology is really an umbrella term that covers a wide spectrum of scientific tools based on molecular biology. As a result it is difficult to define clearly an entity called the 'biotechnology industry'. Rather, biotechnology is a broad, enabling technology that has an impact on productivity in a wide range of sectors (Arunde and Rose, 1998). The agricultural biotechnology community includes dedicated biotechnology firms, established corporations with a biotechnology division, university departments, national and international research institutes, venture capital firms, regional associations, regulatory authorities and suppliers involved directly or indirectly in biotechnology.

Many of the larger agribusiness and life sciences companies have substantial resources and hence are now key players in global agricultural research. For instance, Monsanto has approximately 22 000 employees, has annual R&D investments of around $200 million and generates over $7.5 billion annually in sales. Pioneer HiBred has an annual turnover of US$1.7 billion and invests approximately US$136 million annually in research and development (Joly and Lemarie, 1998). The Mexico-based Empresas de la Moderna's agrobiotechnology division had estimated sales of $572 million in 1995. Private sector investment in specific technologies such as agricultural genomics may outstrip that of most governments. For instance, Novartis is providing a total of $600 million in funding over 10 years to establish an Institute for Functional Genomics in California, USA (Ratner, 1998).

Who funds agricultural research can have major implications regarding the types of technologies that are produced (Buttel, 1986; Lacy *et al.*, 1989; Coffmann and Smith, 1991). Synergistic private–public sector cofunding of research initiatives can be a very cost-efficient means of technology development, provided that the roles and benefits accruing to each partner are balanced and transparent. Some propose that industry funding of cash-starved public sector institutions such as universities or national agricultural research systems (NARS) can in some cases bias research towards the development of input-intensive or other commodity-oriented technology development because input-intensive technologies or products tend to be the most profitable markets for industry (Coffmann and Smith, 1991). There are consumers and crops which do not currently represent a viable market for many companies. Increased long-term public sector funding for agricultural biotechnology which is explicitly aimed at the needs and crops of poorer people will be essential if such people are to benefit from current scientific advances.

Capital Concentration in Agricultural Biotechnology: Mergers and Acquisitions

The past decade has seen a wave of corporate activity in mergers, acquisitions and the creation of new companies in the agricultural biotechnology sector (Kalaitzandonakes, 1998; Lesser, 1998). Seed in the form of elite proprietary varieties has proved to be the delivery mechanism of choice for capturing value from the new input and output traits developed by plant biotechnology research (Lesser, 1998). This has led to high values being placed by agricultural biotechnology companies on 'downstream' seed companies which have high-value portfolios of proprietary varieties and good seed distribution networks. Food processing and distribution companies which are even further 'downstream' may also be rational targets for mutually attractive mergers, acquisitions and strategic alliances with the more 'upstream' agricultural biotechnology and seed companies (Shimoda, 1998). Such mutual attraction may reflect the fact that agricultural inputs only accounted for 8 per cent of the total food industry value in 1992, whereas food manufacturing and retailing accounted for 56 per cent (Cook *et al.*, 1997).

On the other hand, time lags from biotechnology research to product delivery has led to some agrochemical companies (such as Shell, Sanofi and Upjohn) divesting themselves of their seed and biotechnology businesses in the 1990s. The effects on public perception of an intensive campaign of anti-biotechnology lobbying in Europe since the early 1990s, and to a lesser extent worldwide, has led to some companies (for example, in Denmark, Carlsberg and Danisco) withdrawing completely from plant or agricultural biotechnology research (Hodgson, 2000). Most of the recent mergers and acquisitions form part of a broader strategy towards vertical integration of research and development inputs, seed production and distribution channels within commercial crop markets. Some analysts propose that the current restructuring of the global crop seed industry is to some extent based on intellectual property rights portfolios (Sehgal, 1996; Lesser, 1998). However, others contend that patents have only provided weak protection for biotechnology companies with the result that the companies have been forced to engage in vertical mergers and acquisitions in order to better capture value from and protect their technological investments (Kalaitzandonakes, 1998).

As a result of recent mergers and acquisitions, several major players in the agricultural biotechnology sector are now emerging (Lesser, 1998; Moore, 1998). There are now fewer small agricultural biotechnology companies which have not been bought by the ten largest agricultural biotechnology companies (Hayenga, 1998). These mergers and acquisitions have contributed to a restructuring of the seed industry (Lesser, 1998). For instance, having previously owned a 20 per cent stake in Pioneer HiBred, on 15 March 1999 DuPont became the complete owner of Pioneer HiBred in a US$7.7 billion stock and cash acquisition (see *http://www.pioneer.com*). Other agricultural biotechnology mergers and acquisitions include those by Monsanto (which has bought into Agracetus, Agroceres (Brazil), Ecogen, Calgene, Cargill Seeds, Asgrow, DeKalb Genetics and Holdens); AgrEvo (which has bought into Plant Genetic Systems); Empresas La Moderna (which has bought into DNA Plant Technology); Zeneca (which has bought into Mogen International) and Dow Agrosciences (which has bought into Mycogen, Illinois Foundation Seeds) (Joly and Lemarie, 1998).

On 31 March 2000, the former Monsanto Company completed a merger transaction with Pharmacia & Upjohn, Inc. to create Pharmacia Corporation. Another recent merger has been the creation of Aventis Agriculture which unites the crop protection business of Rhône-Poulenc with the crop protection, seeds and crop improvement activities of Hoechst Schering AgrEvo. Another merger has been the establishment of Novartis from a merger of Ciba-Geigy and Sandoz. An interesting subsequent development has been that some of the larger pharmaceutical companies have recently divested themselves of their agricultural biotechnology divisions, in order to concentrate their research focus on the more commercially attractive medical biotechnology market. For instance, Novartis recently divested itself of its agricultural division, which is merging with AstraZeneca's agricultural division to form a new agricultural biotechnology company called Syngenta.

Another feature of commercial strategies over the past decade has been the development of strategic 'partnership' research alliances between agricultural biotechnology companies, especially those with complementary research and patent portfolios. Research synergies can also be achieved through such alliances whereby research capabilities and technology are shared across multiple product lines. Such synergies result in cost reductions and greater potential for new product development (Bjornson, 1998). For instance, Monsanto has established an exclusive research alliance with Millennium (Marshall, 1997) and also with the Mexican multinational Empresas de la Moderna (Massieu, 1998). Similarly, prior to their merger, Pioneer HiBred and DuPont had established a speciality joint research venture (McGowan, 1997). There are also examples where the larger agricultural biotechnology companies are becoming integrated at the shareholder level. Monsanto's acquisitions of 49.9 per cent of Calgene, 45 per cent of Dekalb and 100 per cent of Agracetus involved strategic proprietary technology alliances. Acquisitions of smaller companies such as Plant Genetic Systems, DNA Plant Technology and Mycogen by larger agrochemical companies such as AgrEvo, Empresas de la Moderna and Dow Elanco are considered to have been predicated on obtaining reciprocal access to proprietary biotechnologies. While many of the larger companies are involved in major legal disputes over patent rights and technology contracts, the legal negotiations to resolve such disputes can in some cases coexist with strategic research alliances between the disputing parties (Hayenga, 1998).

Some of the larger agricultural biotechnology companies are also now establishing international joint ventures with both private and public sector institutions in some developing countries (Byerlee and Fischer, 2000; Teng *et al.*, 2000). Successful seed companies in developing countries which have well established seed or planting material distribution systems are also likely to be targets for future mergers and acquisitions (Byerlee and Fischer, 2000). For instance, Monsanto is involved in mergers and acquisitions of seed companies in India (for example MAHYCO, EID Parry), China (for example CASIG, Hebei Provincial Seed Co), Brazil (for example Agroceres, BrasKalb) and Southeast Asia (for example DeKalb) (ibid.). Micropropagation companies which provide farmers with quality planting materials on a commercial basis are also likely to be logical targets for further mergers and acquisitions.

In some countries, antitrust enforcement policies are sometimes required for consumer protection when competition between industries is stifled because particular companies have monopolistic or oligopolistic control of a market. For instance, in some cases of mergers in the pharmaceutical biotechnology industry (such as Hoechst AG's acquisition of Marion Merrell Dow Inc.) the US Federal Trade Commission has intervened in order to preserve competition in the research and development of drugs used to treat medical conditions such as tuberculosis or Crohn's disease. The recent level of amalgamation of the agricultural biotechnology industry has led to antitrust considerations of certain agricultural biotechnology markets (for example, cotton seed and glyphosate herbicide markets) within the USA (Fox, 1998) because it is thought that competition between companies spurs innovation (Hayenga, 1998).

In both developed and developing countries, the current prospects for independent survival for small companies in agricultural biotechnology are bleak because of tortuous regulatory processes (Miller, 1999) and the high risk of incurring patent litigation costs. Many small companies involved in agricultural biotechnology are now assuming the role of service contractors to the larger companies. Indeed, high levels of regulation are increasingly thought to favour the larger companies which have the financial and legal wherewithal necessary to get useful biotechnology-derived products (especially of a transgenic nature) through the tortuous regulatory systems now being constructed. High levels of regulation of the biotechnology sector will favour larger companies and may be intended to act as barriers to entry for smaller companies (ibid.). Indeed, the same barriers to entry may exist for (m)any biotechnology products or technologies developed by the public sector or delivered by public sector mechanisms such as agricultural extension services. In the context of food security, it is therefore worth considering that increased regulations will automatically select for commercial products entering the regulatory systems which are aimed only at the most affluent farmers or lucrative markets. Non-disaggregated overregulation of all agricultural biotechnology could actually widen both the technology and the income gaps between richer and poorer farmers (or consumers).

How commercial considerations now affect publicly funded research is illustrated to some extent by the development by a university research team of transgenic 'Golden' rice which has high levels of vitamin A (Kryder *et al.*, 2000). Golden rice is widely considered to be a biotechnology innovation that could help efforts to combat vitamin A micronutrient deficiency affecting millions of poorer children (Ye *et al.*, 2000; *http://www.micronutrient.org*). Because of the proprietary nature of many of the gene components and the 'enabling technology' processes used to engineer the Golden rice, it was necessary for the developers to negotiate with 32 patent holders of up to 70 patents (mostly held by the private sector, with six being held by universities) before they could begin to disseminate the technology for incorporation into rice varieties used by poorer social groups (Kryder *et al.*, 2000). After negotiations with the patent holders it was agreed that the technology would be available free for non-commercial (humanitarian) use in developing countries. The definition of 'non-commercial' is yearly profit from the Golden rice of less than $10 000 annually. Commercial development will be under licensing from ASTRA Zeneca. Much legal assistance was provided by the goodwill of ASTRA Zeneca to negotiate the licensing exemptions from the other patent holders. Additional regulatory hurdles will probably face the transgenic Golden rice in the form of emerging biosafety regimes. For instance, biosafety considerations have impeded the dissemination of cassava mosaic virus-tolerant transgenic cassava to poorer farmers in Africa (Thro and Spillane, 2000). The current regulatory and commercial landscape is not geared towards the rapid dissemination of useful technologies to poorer groups that could benefit.

Private–Public Sector Research Collaborations

A significant level of agricultural biotechnology research is performed in public (for example NARs) or semi-public (for example universities) institutions and is largely funded by public funds. The broader objectives of publicly funded scientific research differ according to the political economy of different countries. It is commonly accepted that market failures in agricultural research and development lead to underinvestment in research if left solely to the private sector; that is, research opportunities that would be socially profitable go unexploited (Byerlee and Fischer, 2000). The solutions or arrangements proposed for solving the underinvestment problem will largely depend on the type of market failure to be rectified (Alston and Pardey, 1995). For instance, public sector scientific research has often focused on either basic research or on pre-commercial research activities. However, many public sector agricultural research institutions, such as NARs, universities and the CGIAR, play a vital role in conducting applied agricultural research to meet the needs of poorer groups of society, who do not represent a lucrative commercial market for most companies (Tripp and Byerlee, 2000).

In the plant biotechnology sector, publicly funded research may become more commercially driven as a result of shortfalls in public funding, changing incentive structures (such as IPRs and government funding criteria) and increasing research integration between the private and public plant biotechnology sectors. Public sector research institutions may be in direct research competition with private sector firms regarding specific traits and/or clients (Pray *et al.*, 1991). In many instances, public sector research institutions have no clear policies regarding what their comparative research advantages are (that is, how their research objectives differ from those of the private sector) and how their research objectives might complement rather than compete with the private sector (for example, Hossain *et al.*, 2000). Such policies will be dependent on the mandates of public sector institutions; whether they wish to complement domestic and/or foreign private sector research objectives, and the political economy model of dissemination of agricultural technologies to clients. Tripp and Byerlee (2000) have raised a number of questions that policy makers and applied research managers should consider regarding the role of public sector funding and institutions for applied agricultural research (Byerlee and Alex, 1998).

1. If the sale of research products is feasible and profitable, why should the public sector be involved?
2. If the public sector is motivated by financial rewards, will its research be diverted to serve better-off regions and farmers at the expense of small-scale farmers and more marginal areas that might be the primary target of national policy for public research organizations?
3. If a public institution sells non-research products, will research outputs decline?
4. If technologies are protected by intellectual property rights, do the funds generated from research justify the cost of collecting revenues from contracts and royalties?
5. If research generates funds from new technologies, how should the

funds be divided between the programmes responsible for developing the product and the rest of the research organization?

Some public sector plant biotechnology labs now conduct research on particular technology modules on a 'contract' basis for private sector agricultural biotechnology companies. Features of many such research collaborations are confidentiality and non-disclosure clauses – a business approach which is wholly valid in private–private sector alliances but which, in cases of public–private sector alliances, is likely to lead increasingly to queries regarding the rationale of such non-transparency, and possibly declining political support for public funding of the public sector partners. In many instances, such confidentiality requirements may be largely unnecessary and only serve to aggravate suspicions regarding the nature of private–public sector alliances. Open book management strategies may be appropriate for public–private sector alliances in the agricultural biotechnology sector (Case, 1995).

Many leading public sector research institutions are establishing research collaborations with private sector companies. For instance, Novartis has entered into a $25 million deal to fund plant and microbial biology with the University of California at Berkeley, USA (Anon, 1998b). In this instance, Novartis will not have exclusive rights to research findings at the university. Similarly, Zeneca has entered into a $82.5 million research deal with the John Innes Centre (UK) to develop improved strains of wheat over a ten-year period. The French Genoplante initiative is a multi-partner genomics research effort involving the Institut National de la Recherche Agronomique (INRA), the Centre de Coopération Internationale en Recherche Agronomique (CIRAD), the Institut de Recherche pour le Développement (IRD), the Centre National de la Recherche Scientifique (CNRS) and the seed businesses Biogemma, Bioplante and Aventis (Seeker *et al.*, 2000).

Public funding for plant biotechnology research can be contingent upon having commercial research partners, and research success is often partly measured on the basis of what proprietary technologies are developed. For instance, in the USA the 1980 Bayh–Dole Act and the Stevenson–Wydler Technology Innovation Act encourage non-profit organizations such as universities and research institutes to retain certain patent rights in government-sponsored research and encourage the funded entity to transfer the technology commercially to third parties. The Stevenson–Wydler Act requires agencies to establish Offices of Research and Technology Applications at their federal laboratories, and to devote a percentage of their research and development budgets to technology transfer. As a result, some publicly funded US research institutions cannot accept any research materials or technologies which have restrictive patent exclusion clauses.

Access to certain types of EU funding for plant biotechnology research can also be contingent on having commercial collaborators (Lawler *et al.*, 1998). Indeed, in some countries patents are increasingly being considered to be a better indicator of public sector research performance than the more traditional route of peer-reviewed publications (Pavitt, 1997).

The generation of employment through the privatization of publicly funded research is encouraged in the university biotechnology sector through the establishment of spin-off or start-up companies (Horton, 1998; Senker *et al.*, 2000). Many of the more successful plant biotechnology campus-derived companies have now been acquired by the larger agricultural biotechnology companies. Technology transfer accomplishments from publicly funded research to private companies is now increasingly incorporated into the reward and promotion systems of public sector institutions (Meagher and Bolivar, 1998). It is difficult to discern distinctions between institutional incentives which promote short-term commercial gain and those that promote broader social and economic impacts of research, especially in relation to food or livelihood security.

When it comes to the needs of poorer farmers and consumers, very few public sector agricultural research institutes or funding mechanisms have incentive systems which reward those who meet the needs of clients and hence make research staff more accountable to their clients (Collion and Rondot, 1998). For the majority of public sector scientists involved in plant biotechnology research, their reward system is largely dependent upon publications. Yet, when it comes to questions of research for food security-related objectives, additional indicators of research success could also be factored into public funding mechanisms. For instance, adoption rates of plant varieties by farmers and other types of indicators of client satisfaction with the products of crop improvement research may be valid but underutilized research variables (Farrington, 1994). Technology adoption rates and adoption quality could be used in much more innovative ways to construct reward systems for both public and private sector scientists (Farrington, 1997).

An increasing number of universities in the OECD countries now manage patent portfolios based on their research from which they try to generate revenue through licensing. In most instances, the revenues generated are relatively small and often only cover the costs of the university intellectual property technology transfer office. Surplus licensing revenues are typically reinvested in research activities. Licensing of patents generated from public funds can be on the basis of either exclusivity or non-exclusivity. While exclusive rights to a patent are likely to be more expensive and hence generate more revenue for public sector institutions, significant revenues can also be generated from non-exclusive licences provided to a wider range of licencees at minimal fees. In the biotechnology sector, one example of successful non-exclusive licensing is the exceptional $38.5 million generated in 1997 for the US universities which hold key patents on genetic engineering technology (Horton, 1998). If patents on research which is entirely publicly funded were limited to non-exclusive licensing provisions (for example as a precondition of public funding) this would be likely to increase competition between companies and other research entities, and could help to ensure broader access to useful biotechnologies generated as a result of public funding.

A number of problems are typical of public sector involvement in IPR protection. These include overeagerness of university technology transfer managers to file patent applications, their overestimation of the value of their intellectual property and the underestimation of the additional investment required to turn a research discovery into a product, and their readiness to grant exclusive, rather than non-exclusive, licences. There is a need for governments to make their criteria for provision of public funding for agricultural biotechnology research more specific and transparent regarding IPRs, and also to identify more clearly who the primary clients of such research funding are.

Technology transfer lawyers in research institutions operate primarily within the boundaries of their respective national laws. Hence national legislation or conditionalities regarding public funding of agricultural research can be used as a policy instrument to ensure that broader social and economic goals are met through publicly funded biotechnology research. In addition, the encouragement of research institutions to develop clear mission statements and correspondingly transparent IPR policies could help to broaden access of poorer social groups and countries to useful biotechnologies. For instance, the provisions of the Convention on Biological Diversity on biotechnology transfer to developing countries are rarely considered by many public sector agricultural biotechnology research institutions. However, there are a number of provisions of the Convention on Biological Diversity which are relevant to the transfer of biotechnologies between research institutions in different countries. These include potential rights of the biological resource donor country's institutions:

- participation in research, Article 15(6),
- sharing in the results of research and proceeds of commercial exploitation, Article 15(7), and
- access to and transfer of the derived technology, Article 16(1).

Some types of public sector institutions (such as universities) lack the capacity to both research and develop a product to its final stage. Public sector plant biotechnology research in the OECD countries has become increasingly 'atomized', whereby each laboratory may develop and patent a vital modular component of a product but is rarely directly involved in developing the final technology package embodied in novel plant biotechnology products. The final products of plant biotechnology research in the OECD countries are typically distributed by the private sector. However, there are numerous examples of publicly funded breeding programmes, seed production and agricultural extension in developing countries that also try to develop and provide finished products (that is, varieties) to poorer farmers and consumers (Byerlee and Fischer, 2000).

An increasing number of final biotechnology-derived products are composed of a range of proprietary modular components that are often owned by a number of different parties. As a result, public sector institutions (such as NARs or CGIAR) which are involved in providing finished products (such as seed or varieties) to farmers and which wish to incorporate useful proprietary traits in their varieties will increasingly have to negotiate terms of legal access to such inventions with the patent owners. The Third CGIAR System Review recognized this and recommended that the CGIAR establish a legal entity which could hold patents, and develop rules of engagement (involving both the public and private sector) based on the premise that access to the means of food production is as much a human right as access to food (see *http://cgreview.worldbank.org/cgrevrep.html/*).

Public sector funding bodies could also play a stronger role in specifically directing plant biotechnology research towards longer-term social and economic objectives rather than short-term commercial objectives. In this respect, it will be very useful if the private sector can transparently identify those public goods in the agricultural biotechnology arena which it cannot provide or fund in the short to medium term. At present, the International Seed Trade Federation (FIS) states that, up to now, fundamental research done by the state institutes has in a great number of countries largely contributed to and stimulated the work done by private companies. It is felt by FIS that there will be an increasing need for fundamental research. However, FIS is convinced that farmers will benefit best if the results of this basic research are developed, increased and distributed by private companies (see *http://www.worldseed.org/~assinsel/pos_fis.htm*).

The needs, constraints and objectives facing biotechnology researchers in the public and private sectors can differ widely, yet public sector scientists who conduct agricultural biotechnology research have not been well represented or very actively involved in policy formulation at the national or international level. In most instances, the views of the private sector plant biotechnology industry, on the one hand, and environmental organizations concerned about aspects of agricultural biotechnology, on the other hand, have had the most impact on the formulation of national and international biotechnology-related policy. Neither of these interest groups is necessarily representative of publicly funded plant biotechnologists, agricultural researchers, or consumers in general.

There has been a significant lack of policy input from public sector scientists as a group. For instance, there has not been much involvement in intergovernmental fora (such as the Convention on Biological Diversity meetings, the FAO Commission on Genetic Resources, the World Trade Organisation or the Codex Alimentarius) or membership-based international or regional scientific NGOs (for example, the Third World Academy of Science, the International Society for Plant Molecular Biology, the International Association for Plant Tissue Culture, the African Biosciences and Plant Biotechnology Networks, the International Society for Tropical Root Crops, the Cassava Biotechnology Network or the African Association for Biological Nitrogen Fixation) which mainly represent public sector biotechnologists and agricultural research scientists. Most such membership based-groups do not have full-time representatives who are professionally trained to lobby on their behalf. As a result, most policy-related interventions from such groups are typically in the form of periodic statements (reactive and made once, rather than proactive and sustained) which are unlikely to be the best communication strategies for promoting policy change.

Orphan Crops and Orphan Drug Acts

Plant biotechnology to date has been biased towards some crops rather than others (James, 1997). Privately funded plant biotechnology research is heavily biased towards major commercial (often export) crops such as maize, soybean, canola, cotton, tobacco, tomato, potato, squash and papaya, which are the main species for which commercial transgenics have been promoted on a large scale. This has been driven by the cost of developing biotechnology-derived crops and the potential large markets for a relatively small number of commodities such as maize and soybean. Similar crop biases are also evident in the crop focus of public sector plant biotechnology research in many OECD countries.

By comparison, plant biotechnology research on many of the 'orphan' crops of poorer peoples such as millets, yams, plantains, cassava and sweet potato is limited in terms both of its research intensity and of levels of public sector funding. A similar situation exists for the many so-called 'Cinderella' trees and shrubs of major importance to poor people's livelihoods which have been overlooked by agricultural researchers. The private sector has little interest in investing in the improvement of such crops, unless there is scope for increasing sales of such crops in global commodity markets such as starches, oils or animal feed. Even within particular crops, there may also be a level of varietal bias in current plant biotechnological research. There may be important differences between varieties regarding their response to tissue culture or whether they are easily genetically transformed. A significant effort is often needed to develop regeneration and transformation protocols for specific varieties with a concomitant bias towards commercial or export varieties. For instance, there is a current need for the development of transformation protocols for the different rice varieties which are associated with different rice agro-ecosystems.

In the area of biomedical research, the lack of commercial incentive for companies to develop therapeutics for diseases with a small number of sufferers (such as rare diseases) or with large numbers of financially poor sufferers (such as shistosomiasis or malaria) has been recognized for many years. Government legislative intervention in the form of 'orphan drug acts' has been established in the USA (1983, *www.fda.gov/orphan/index.htm*) and Japan (1985), and are proposed for the European Union (see *http://www.etm.nl/Newsletters/News 98/Autumn/page3.htm*) to provide incentives for private sector research into therapeutics for rare diseases (Davidson, 1996). Incentives under such orphan drug acts can take many forms, such as:

- limited period market exclusivity,
- grants,
- tax credits,
- regulatory assistance,
- clinical trials assistance,
- subsidies,
- preferential access to public sector research funding,
- 'fast track' regulatory trials.

Some existing orphan drug acts have been criticized on a number of issues (Love, 1999). It is questioned whether the existing acts can function to promote research on the development of unprofitable drugs, for instance where the end-users are too poor to warrant any private investment. It has also been suggested that the market exclusivity provisions in orphan drug acts, which can be broader in scope than patent protection, can be used to create barriers to entry by other competitors. More rigid criteria for designation of orphan drugs, selective incentives for research and greater public accountability are considered to be remedies for the better functioning of orphan drug acts (ibid.).

Orphan drug acts are in essence limited period joint ventures between the public and private sectors to meet the need for some public goods. Similar arguments can be made for public/private sector cofinancing of research on orphan crops of smallholders, in the same way as state/commercial interests coincide in the orphan drug acts, which promotes research in drug development for groups of sufferers who do not represent a large enough commercial market for total private sector interest. For some crops, there may be a need at both national and international level for the development of analogous 'orphan crop acts' to stimulate research and development for locally consumed orphan crops such as cassava, yams and millets. It is likely that most export crops will continue to attract sufficient private sector investment.

Because the basic biotechnology research tools for commercial crops such as maize, wheat, soybean and cotton could equally be applied to orphan crops, public sector incentives could tip the balance towards increased public and private sector investment in research to meet the needs of consumers of orphan crops. However, this would require the adequate definition of what an orphan crop is and what conditions limit private sector investment in research on the crop in question. In this respect the development of orphan crop acts could learn from the experiences gained in the development of orphan drug acts. Countries who wish to stimulate private sector investment in orphan crops could consider that 'orphan crop acts' with limited duration exclusivity and sales caps may be one means of stimulating research on crops of importance to national food security. Such sales caps might easily be determined from international export figures. This may provide a strategy for the CGIAR and some NARs to get preferential access to proprietary technology for some of their mandate orphan crops which are not internationally traded in a significant commercial sense (Binenbaum and Wright, 1998).

Social Venture Capital: Developing New 'Push–Pull' Funding Strategies for International 'Public Good' Agricultural Biotechnology Research

Donor efforts in agricultural biotechnology are highly fragmented and projectized (Thro and Spillane, 2000), with no multi-donor consensus on priorities for co-ordinated multilateral research to meet the needs of poorer farmers and consumers (Byerlee and Fischer, 2000). Political vacillation in many industrialized countries regarding the role of agricultural biotechnology in food surplus-prone agricultural economies is also having negative impacts on decision making regarding the role of agricultural biotechnology in agriculture in food deficit countries, especially among the OECD donor community. In such a research funding climate, new approaches to the funding and management of agricultural biotechnology research to address 'public good' problems which are international in nature are needed (Pinstrup-Anderson and Cohen, 2000).

The traditional type of funding, where scientists apply for a grant and then conduct their applied research, is called 'push' funding. One problem with this approach is that scientists are encouraged to exaggerate their chances of success, and often final products never materialize or reach the intended beneficiaries. For instance, effective vaccines against hepatitis B and Hemophilus influenza type B were developed over a decade ago and are now only beginning to become available in developing countries. Push mechanisms are typically in the form of direct or indirect financial support that offers incentives to perform particular types of research (Gill and Carney, 1999).

A complement to the 'push' funding model is 'pull' funding, whereby a research incentive fund is established to reward research that develops a 'finished' technology or product that meets certain agreed-upon criteria. In pull funding, there is no 'public venture capital' funding until the finished product is developed. Pull mechanisms are expected to encourage larger companies to make long-term investments in 'public good'-type research by ensuring that such 'public good'-type research will result in an adequate profit. Pull funding is considered to have potential for creating incentives for industry to invest in research and development for developing market products (for example, vaccines for poorer social groups) with a high volume/low profit margin.

A consortium of funding agencies (donors and so on) could establish separate international funds for the development of specific technologies to address a specific problem in developing countries (Sachs, 1999). In the agriculture sector, international competitive funds could be established to bid for the supply of key technologies or products. Tenders from research organizations both in the public and in the private sector could be considered to identify the most cost-effective and high-likelihood approaches to the development of a specific technology or product. The funding consortium could place conditionalities on the ownership of the products of the research, if for instance they wanted to ensure that the research products would be widely accessible.

One current example of such an approach to technology development is the multi-partner Global Alliance for Vaccines and Immunisation (GAVI) coalition (WHO, 2000) dedicated to encouraging the expanded availability and development of vaccines for developing countries (see *http://www.vaccinealliance.org/*). In addition to providing major incentives for the practical delivery of existing vaccines to beneficiaries in developing countries, the GAVI initiative intends to create a US$1 billion fund to promote research into effective and accessible vaccines for HIV, malaria and tuberculosis. The GAVI initiative also intends to assess and set priorities for which developing-market vaccines (such as cholera, typhoid fever, dengue fever, leishmaniasis and shistosomiasis), in addition to HIV, malaria and tuberculosis, are most needed. The GAVI initiative recognizes that the costs of patent filing for vaccine-related research in public sector institutions may be prohibitive, and is exploring approaches whereby GAVI would provide IP protection and management services for the public sector partners.

A related approach to create symbiotic public–private partnerships specifically to develop affordable technologies for developing countries is for the public sector to establish a not-for-profit (NFP) company which is dedicated to fostering development of a particular technology or product in a manner accessible to poorer social groups in developing countries. The International AIDS Vaccine Initiative (IAVI) serves as a model of such a virtual corporation dedicated to the development of a specific technology (that is, an AIDS vaccine) by accelerating scientific progress through contracting research services from both the public and the private sectors (see *http://www.iavi.org*). Since 1996, IAVI has invested $20 million in a series of international vaccine development partnerships specifically aimed at HIV strains in developing countries.

In addition, IAVI has negotiated intellectual property agreements with industry partners to help ensure that vaccines will be readily available in developing countries at reasonable prices. IAVI considers that its investment of 'social venture capital' enhances the value of its industrial partners' intellectual property, without interfering with their most profitable markets. The World Bank, G-8 leaders, EC and other international partners are working with IAVI to establish Vaccine Development and Purchase Funds, financial instruments to encourage the commercial sector's investment in developing AIDS vaccines for developing countries. A Vaccine Purchase Fund is anticipated to create a guaranteed paying market of known minimum size in the developing world. There is much unexplored scope for parallel approaches to utilizing the complementary agricultural biotechnology capacity of the public and private sectors to meet needs of poorer social groups.

INTELLECTUAL PROPERTY RIGHTS AND RELATED DEVELOPMENTS

Intellectual property rights (IPRs) represent a useful means by which private investment in research and development can be promoted. IPRs provide commercial incentives for research and development activities by prohibiting direct copying without permission (for example, the payment of a royalty). The concept is that the inventor or other creator cannot compete with a copier who shares none of the development costs. In return for the risk of such investment, the IPR owner obtains for a limited period (say 20 years) the right to use the intellectual property exclusively, assign ownership, licence it or not use it at all. The IPR owner can enforce these property rights if others misappropriate or otherwise infringe the protected intellectual property. In a social sense intellectual property rights represent a means to promote commercially relevant innovation, and are not an end in themselves (Leisinger, 1996).

When considering the incentive effects, it is important to recognize what privileges IPR do and do not provide. They do not assure a return; indeed, only up to 15 per cent of patents are ever commercialized. Hence predicting the level of future use from the simple act of filing a patent is a difficult if not impossible exercise. IPRs do not necessarily permit the use/practice of the creation, as this is often controlled by other regulations (such as on biosafety) or even other patents. Primarily, they allow the right to exclude others from use: what can be called 'negative rights'. All financial rewards must typically come from market sales, although social rewards might accrue from licensing at lower rates to non-commercial users. For most IPRs, key factors such as the breadth (scope) of protection are critical in determining the commercial value of the IPR. IPR legislation is national law, applying only in those countries where it is available and has been granted.

It is not yet clear whether IPRs are suitable incentives for public sector biotechnology research for food security and other public goods (Barton and Strauss, 2000). For IPRs generated by both the public and the private sectors, this will depend on whether such IPRs are practised for narrow commercial objectives, or whether broader social or economic objectives are factored into promoting the licensing and use of IPR-protected biotechnologies. In any market-driven system with strong forms of IPRs, it is to be expected that investments will be aimed at projects with the greatest commercial rate of return. While IPRs are excellent incentives to stimulate private sector innovation for commercial gain, they potentially have a distorting effect on the research objectives and directions of public and semipublic sector institutions. The application of IPRs as research incentives for publicly funded research may tend to reorient research towards short-term commercial objectives rather than longer-term economic gain for the 'public good'. In essence, the same IPR incentive structures can lead to competition between the research objectives of the private and public sectors for the same commercial markets. In such instances, the needs of resource-poor farmers who are not a viable commercial market for the private sector are likely to continue to be unmet.

In the agricultural research arena, both scientific knowledge and its commercial applications are increasingly becoming proprietary. Proprietary rights over agricultural biotechnology products and processes are being claimed by both private firms and an increasing number of public institutions. Such rights include trade secrets, patent rights, plant varietal protection (PVP) and contractual rights arising from the use of material transfer agreements (MTAs) (TAC, 1998). How regulations, policies and incentives regarding public sector funding of agricultural biotechnology research are framed by governments will have a major effect on whether agricultural biotechnologies can have a beneficial effect upon global food and livelihood security. In particular, it will be increasingly important for governments to identify more specifically who the primary beneficiaries of public research funding are, what the purposes of such funding are and which stakeholders will have real access to any useful biotechnologies generated from public funds. In this respect, a clarification of the complementary roles and objectives of the domestic private and public sector institutions will inform the development of better incentive structures to promote both commercial and economic development. While IPRs may be the best incentives for private sector institutions, alternative or modified incentive structures (such as limitations on exclusive licensing) may be more appropriate for public sector research institutions (or for publicly funded research).

At the international level, the Uruguay Round of the General Agreement on Tariffs and Trade (GATT) has created a subsidiary agreement on Trade Related Aspects of Intellectual Property Rights (TRIPs). Any country ratifying GATT accepts the obligation to establish minimum standards of intellectual property. The TRIPS Agreement was reviewed in 1999. The current revision of the WTO's TRIPs Agreement is therefore of importance to future trends and research directions in agricultural biotechnology research. The current coexistence of different models of IPR law between different countries and regions (see below) will probably lead to negotiations for international harmonization of IPRs within the WTO.

The majority of patents and PVP certificates currently filed are filed by companies predominantly from the OECD countries. It is unclear at present what impact the harmonization of IPR systems will have on the relative roles of foreign and domestic innovation in the agricultural biotechnology sector in developing counties. Indeed, this will to a certain extent depend upon what IPR models result from any international harmonization of IPRs. As part of the Transatlantic Economic Partnership, the European Commission and the USA have reopened negotiations on the harmonization of US patent law with that of other countries (Masood, 1999). The CGIAR's Third System Review has stated that negotiations regarding agriculture and intellectual property rights within the WTO will have major bearings on global food security and in shaping the context within which public sector research institutes will have to operate.

Patents

National patent laws provide protection of inventions demonstrating the key characteristics of novelty, non-obviousness, utility and sufficient disclosure. A patent confers a right on the patent owner to prevent others from freely exploiting what is claimed in the patent. The patent system applies generically to stimulate private sector investment in research and development in many sectors, of which agriculture is but one.

There are significant differences between the patent procedures of different countries and regions regarding agricultural biotechnology (Agris, 1998a). For instance, while most of the world's patents laws operate on the 'first to file' basis, there are some countries such as the USA which operate on the 'first to invent' basis. The US patent system allows prior publication for up to a year before filing while most other countries regard prior publication as prior disclosure. Most patent regimes require that biological material involved in the patent be deposited in a germplasm collection either before filing (most countries) or after issuance (for example, USA). The options for intellectual property rights protection for plants also differ considerably between the USA and the EU (Agris, 1999). In the USA, utility patents and plant patents can be taken out on the entire genotype of some types of plant varieties (such as mutants and asexually reproducing plants), whereas Article 53(*b*) of the European Patent Convention states that a plant variety or biological process for the production of the plant variety is unpatentable.

Another difference between the European patent system and US patent law is that the term 'invention' means invention or discovery in the US system. In European law 'discovery' is distinguished from 'invention' and is considered unpatentable. However, the distinction is not easy to define. A discovery involves new knowledge whereas an invention is a practical application of knowledge. Naturally occurring substances present as components of complex mixtures of natural origin can in principle be patented where they are isolated from their natural surroundings, identified and made available for the first time and a process is developed for producing them so that they can be put to a useful purpose. This applies to inanimate substances as well as to living materials. In some circumstances such substances are not ruled out as mere discoveries but are considered as inventions. Microorganism patents are granted by the US, European and Japanese patent offices.

While patent regimes differ internationally, a certain level of international harmonization of IPR law is under way under the WTO Trade

Related Intellectual Property Rights (TRIPs) Agreement. In addition there are a range of international agreements regarding patents. The Patent Cooperation Treaty and the Paris Convention for the Protection of Industrial property are two relevant examples of such international policy. If an exported patent is filed in a country that subscribes to the Paris Convention (most countries) the foreign filed application will be treated as if it had been filed simultaneously with the original application. Hence the expiry date for a single patent should be the same if filed at different times in multiple countries. The International Patent Cooperation Treaty offers a centralized system for filing patent applications for its member nations. Membership is open to any countries that are signatories to the Paris Convention. As of 15 January 1998, there were 95 member nations which were signatories to the Paris Convention. The World Intellectual Property Organization (WIPO) is the United Nations body which is responsible for international aspects of IPRs. There are also a range of regional patent treaties such as the African Intellectual Property Organization (OAPI), the African Regional Intellectual Property Organization (ARIPO), the Eurasian Patent Convention and the European Patent Convention (EPC).

In apparent contrast to many other types of industries, patents are considered as key assets in the agricultural biotechnology industry. Studies of widely different types of businesses have reported that, for some businesses, patents were not a very important means of securing competitive advantages from new products (Levin *et al.*, 1987). However, within the same studies it was found that the pharmaceutical, chemistry, and plastic materials industries did consider patents to be an effective means of protecting new products. In the biotechnology industry, a patent can be viewed as a means of protecting the large, 'up-front' investments necessary to the research and development of new drugs and biotechnologies.

Plant Varietal Protection (PVP) and Other IPR Systems

Plant breeders' rights (PBR) or plant varietal protection (PVP) systems are synonymous terms to describe specialized (*sui generis*) IPR systems for cultivated plants. PBRs were first systematized under the International Union for the Protection of New Varieties of Plants (UPOV) (see *http://www.upov.org*). UPOV is an intergovernmental organization which was established under the 1961 UPOV Convention signed by its member governments. The purpose of the UPOV Convention is to ensure that the member states acknowledge the achievements of breeders of new plant varieties, by making available to them an exclusive property right, on the basis of a set of uniform and defined principles. The UPOV Convention entered into force in 1968 and was revised in 1972, 1978 and 1991. The 1991

Act of the UPOV Convention entered into force on 24 April 1998. Membership of UPOV, among other steps, requires that signatories adopt national legislation along the lines of the 1978 or 1991 (such as USA) UPOV Conventions. As of 23 March 1999, 39 governments had become member states of UPOV. These were Argentina, Australia, Austria, Belgium, Bulgaria, Canada, Chile, China, Colombia, Czech Republic, Denmark, Ecuador, Finland, France, Germany, Hungary, Ireland, Israel, Italy, Japan, Mexico, Norway, Netherlands, New Zealand, Paraguay, Poland, Portugal, Republic of Moldova, Russian Federation, South Africa, Spain, Slovakia, Sweden, Switzerland, Trinidad and Tobago, Ukraine, United Kingdom, United States of America and Uruguay. As of 22 January 1999, 11 out of 39 member states had developed legislation in line with the UPOV 1991 Convention, while the remainder had legislation in line with the UPOV 1978 Convention. The UPOV 1991-compliant countries were Bulgaria, Denmark, Germany, Israel, Japan, Netherlands, Republic of Moldova, Russian Federation, Sweden, United Kingdom and United States of America.

In place of the novelty, non-obviousness and utility requirements of patent law, PBR uses the requirements of novelty, distinctness, sufficient uniformity and stability (DUS). Uniformity and stability are measures of reproducibility true-to-form, respectively among specimens within a planting and intergenerationally. The principal test then is distinctness to determine novelty; that is, that the variety be 'clearly distinguishable from all' known varieties. PBRs differ from patents in a number of key respects and it is generally not useful in discussions of IPRs to confuse or conflate these two different IPR systems.

It is worth noting also that both trademarks and trade secrets are widely used intellectual property rights in the agricultural biotechnology sector. For instance, trademarks can be associated with a particular company name (such as Pioneer Hi-Bred), or individual products like the FlavrSavr tomato. Note that the FlavrSavr tomato genotype may also be patented, so the two forms of IPR can be complementary. Within agriculture, the parent inbred lines used to generate F-1 hybrids may be considered a form of trade secret. So long as the crosses and/or the inbred lines are protected, the product is difficult to copy. However, the self-reproducible nature of most living organisms precludes a major role for trade secrets as IPR protection systems for agricultural products. In other technological areas, trade secrets may substitute for or complement patents. When a product or process is difficult to copy, trade secrets can be a substitute for patents.

The vast majority of patents in plant biotechnology research are taken out by private or public sector institutions in OECD countries (for example, USA, Japan, EU countries, Australia). The same trend is evident for plant varietal protection or plant breeders' rights certificates. It would seem that domestic innovation in plant biotechnology research in the majority of non-OECD countries has to date yielded few IPR-protected technologies or products.

Technology Use Fees and Contracts

Intellectual property rights (IPRs) are not the only legal means that can be used to ensure that proprietary technologies are not misappropriated. Some companies also use bilateral legal contracts with growers to ensure that their products are grown in a particular manner and to ensure that value is captured from such downstream end-users (Renkoski, 1998). Agricultural biotechnology which focuses on output traits will result to some extent in a shift from agricultural production of commodities towards more specialized production for lucrative niche markets.

Agribusiness projections suggest that, by 2028, farmers will be responsible for 10 per cent of the added value in end-products, whereas food processing and distribution will be responsible for over 80 per cent (Goldberg, 1999). This contrasts with relative value added contributions of 32 per cent and 50 per cent in 1950 for farmers and food processing/distribution, respectively. It is thought that an increasingly large percentage of varieties containing such specialized output traits will be produced under strict contractual guidelines (Kindinger, 1998). It is further suggested that such contracts are likely to be managed by input suppliers who, in partnership with the farmer, will produce within contract parameters for specific niche markets (Kindinger, 1998). A number of companies are using such types of legal contracts with farmers who act as contract growers (Freiberg, 1997). In some cases farmers buying such proprietary seed have to sign contracts guaranteeing no reuse of seed in the following year (Hayenga, 1998). Such contracts can also require that the growers use particular brands of inputs (such as herbicides) on the proprietary seed varieties.

Another development has been that of technology fees. To recapture private investments in plant biotechnology research some companies now make a legal distinction between the value of the original seed or variety per se and the value of the new technology embodied in the improved seed. Such companies now charge a 'technology premium' to farmers when they purchase improved seed. For example, the 1996 technology premium for *Bt*-based insect protection in cotton was reported to be about $75 per hectare and $25 per hectare for maize. More recently, the technology premium list prices ranged from $32 per acre for *Bt* cotton to $5 per unit for herbicide-resistant soybeans. In some cases, such technology fees have been waived to facilitate market entry and it is expected that such fees will decline as more competitors bring substitute technologies to the market (Hayenga, 1998).

The developers of novel biotechnologically generated foods may also opt to internalize the entire production and distribution process to the exclusion of other producers and suppliers. Calgene, owner of the FlavrSavr tomato, for example, were reported to be producing exclusively under contract or using their own facilities. Where labour costs are lower in developing countries it is possible that contract growing of proprietary varieties containing output traits (for example, especially for export crops) will begin to emerge.

Technology Protection Systems

There has been some controversy generated over the development of 'technology protection systems' (TPS) or 'genetic use restriction technologies' (GURTs) which aim to ensure that saved seed containing proprietary technologies or genes is not replanted without adequate payment for the novel embodied technologies. Such systems are likely to work for all self-pollinated and outcrossing seed propagated crops, and may have distinct biosafety benefits. At present no such systems have been developed for clonally propagated crops. Such systems may promote private investment in crop research for crops where the extent of replanting of saved seed is a disincentive for recouping investments. One such technology, dubbed the 'terminator', which was originally developed by the USDA and Delta & Pine, and is now owned by Monsanto, has been the focus of much attention. A number of countries (such as India) are reported to have banned the use of such TPS systems and the CGIAR has stated that such systems will not be used in its research programmes.

None of the technology protection systems under development has yet been commercialized (that is, reached farmers' fields) and they are unlikely to be commercialized in the near future (say 5–10 years). If developed further, the next generation of systems are likely to involve the failure of the proprietary genes to express their useful traits when the variety is replanted, rather than the failure of the entire varietal genotype to replicate itself. The 'switching on' of such proprietary traits which rest upon the varietal platform may be contingent upon the application of proprietary chemicals which induce the proprietary transgenes to express the useful trait. Hence farmers could choose whether or not to buy the proprietary chemicals which switch on the improved traits. The impact of such GURTs has been the subject of an independent international expert review process commissioned by the Parties to the Convention on Biological Diversity.

Research Exemptions under Patent and PVP Systems

A key consideration for researchers, and hence for the public concerned about the efficiency of agricultural research, is access to IPR-protected materials and technologies for research or non-commercial purposes. An IPR research exemption refers to the permissible use of protected materials for certain research and product development/improvement purposes, such as non-commercial use. Research exemptions under IPR systems are critical to future scientific innovation and competitiveness. However, there are signs that research exemptions under both patent and PVP systems are becoming less standard and more restrictive in the favour of the IPR holders.

Most patent legislation provides an exemption for scientific research or non-commercial uses of the patented technology. However, the practical nature of such research exemptions can differ significantly between different patent systems. For instance, the nature of the research exemption under the US and Canadian patent systems is more restrictive than the research exemption provided under the European and Japanese patent systems. In some cases the research exemption is explicitly stated in national patent laws, while in others it is not. Furthermore, there can be significant legal differences between experimenting on a patented invention – that is, using it to study its underlying technology and invent around the patent, which is what the exemption covers – and experimenting with a patented invention to study something else, which is not covered by the exemption. Many researchers in the USA and Canada work on the assumption that patent holders are unlikely to file a lawsuit against an academic researcher whose use of their invention is commercially insignificant (Eisenberg, 1997).

The research exemption under most PVP systems (such as UPOV) is typically called the breeders' exemption. The breeders' exemption refers to the right to use protected materials as the basis for developing a new distinct variety or other research use. Research or experimentation exemptions under patents are not as well defined. However, to prevent copying of research, the breeders' exemption under PVP laws such as promoted by UPOV 1991 is now contingent upon the questions of essential derivation and dependency. The 1991 Act of the UPOV Convention requires that varieties eligible for PVP shall not be essentially derived from other protected varieties or require the repeated use of the protected variety (for example, inbred lines for F1 hybrid production). Essentially derived varieties may be obtained for example by selection of natural or induced mutants, by selection of somaclonal variants, by backcrossing or by genetic engineering.

The definition, either by molecular or phenotypic means, of thresholds which would define what extent of essential derivation and/or dependency would constitute an infringement is currently under discussion in UPOV, The International Association of Plant Breeders for the Protection of Plant Varieties (ASSINSEL) and FIS. It is likely that molecular profiling of varieties will be increasingly used to identify varieties which are essentially derived from commercially valuable elite germplasm. Some seed companies (such as Pioneer Hi-Bred) are now hiring germplasm security officers trained in molecular diagnostics who will specialize in identification and prosecution of essential derivation situations where proprietary germplasm is being misappropriated. Unlike the case of state-promoted DNA forensics for identification of humans in criminal situations, it is likely that the DNA forensics will be performed by the companies themselves rather than by independent agencies. It is also likely that criteria for judging what constitutes an 'essentially derived' variety will have to be defined on a crop-by-crop basis.

In the area of IPR protection for plants, utility patents are considered to be the IPR of choice in situations where the technology holder would like the strongest protection and the most minimal research exemption (Agris, 1999). However, the legal interface between plant breeders' rights and patents is currently unclear in many countries. Because plant molecular biotechnology is conducted in a modular fashion, a single variety (or a transgene) can be subject to a multitude of different patents, each on different genetic components or processes, and each possibly owned by different owners in different countries. In addition, such varieties can also be subject to PVP and the legal provisions therein. ASSINSEL has stated that patented plant genetic components, traits or characteristics and commercialized varieties, including their patented genetic components, traits or characteristics, should be unrestrictedly accessible and/or usable for developing new plant varieties. However, to commercialize a variety incorporating a patented genetic component or expressing a patented trait or characteristic, ASSINSEL recognizes that the authorization should be requested from the patent holder (see *http://www.worldseed.org/~assinsel/pos_fis.htm*).

Patent lawyers are typically paid to submit claims for IPR protection which are as strong and broad as possible (Agris, 1998b). There has been much debate on the issue of agricultural biotechnology patents which are increasingly broad in scope (Thayer, 1995). In particular, when broad patents, or patents on basic research/enabling technologies occur there can be a tendency for the patent holders to engage in cross-licensing or patent pooling. Such practices can act as significant barriers to market entry for competitors and can act as disincentives for follow-on innovation (FTC Report).

Some of the larger companies with powerful patent portfolios on agricultural biotechnologies are now involved in cross-licensing or bartering of one patented technology for another. By comparison, isolated public sector research institutions or smaller companies with relatively weak patent portfolios and legal expertise are not in a strong bargaining position to gain access to many useful proprietary technologies under standard research exemption clauses (Ochave, 1997). Threats of litigation or 'sham litigation' to slow down competitive entry into markets will also be disproportionately felt by weaker institutions or organizations (FTC Report).

In the area of agricultural research, there is a greater need for more detailing of valid criteria for research exemptions under national patent law to ensure that researchers can gain access to proprietary technologies for research and non-commercial purposes. In the context of food security in developing countries, there may be some scope for obtaining research exemptions on use of proprietary technologies for non-commercial purposes, such as orphan and neglected crops, non-export crops, subsistence farmers and marginal areas (Byerlee and Fischer, 2000).

A report issued in July 2000 by seven national academies of sciences urged companies to licence their proprietary technologies at concessional rates, for application by the poor in developing countries. There are indications that the major agricultural biotechnology companies are willing to licence some of their proprietary technologies for non-commercial or humanitarian purposes in developing countries. For instance, Monsanto recently stated that it will provide royalty-free licences for all of its technologies that can help further vitamin A-enhanced rice variety development in developing countries, in addition to its public disclosure of its rice genome sequence database (*www.riceresearch.org*). To date, all such initiatives are on a technology-by-technology basis. The impact on incentivating pro-poor biotech research would be much greater if such exemptions (or guidelines for obtaining such exemptions) could be on a company-by-company basis, where companies might consider that proprietary technologies which are on the market (or near market) could be applied in developing countries for non-commercial purposes.

A more transparent elaboration of the research exemption criteria for non-IPR holders coupled to a greater specification of the utility and enablement doctrines for patent holders could increase incentives for follow-on and incremental research and help to deter anti-competitive cross-licensing schemes (FTC Report). In particular, stronger research exemptions or compulsory licensing might be sought in situations whereby the patent owner 'fails to practise' the patent in certain unprofitable 'public good'-type situations, such as for improvement of non-commercial subsistence crops or for less lucrative markets (Barton and Strauss, 2000). Generic material transfer agreements (MTAs) concerning genetic resources or technologies could also be developed which would broaden access to germplasm or technologies for certain purposes (non-commercial use, non-export crops and so on).

The relative lack of participation of membership-based NGOs representing public sector scientists as observers in international fora concerning food security, environment, agriculture and genetic resources has meant that important issues regarding research exemptions and access to technologies remain off policy agendas. There is a pressing need at the national or international level for greater involvement of membership-based organizations representing public sector scientists in the elaboration of criteria and approaches for maintaining IPR research exemptions in a manner which best promotes research competition and equitable technology transfer.

Farmers' Privilege to Save and Resow Proprietary Planting Materials under Patent and PVP Systems

The farmers' privilege is the right to hold PVP-protected germplasm as a seed source for subsequent seasons (farmer-saved seed). Where the privilege exists, farmers can use the harvested product of a protected variety for propagating purposes on their own holdings, where the harvested product was obtained by previous planting on their own holdings. Under Article 15 of the UPOV 1991 Convention, the farmers' privilege is optional, within reasonable limits, for governments to include in their national PVP legislation.

The farmers' privilege to resow saved seed would generally be an infringement with most patented materials. Under most current patent systems, there is no farmers' privilege to allow the saving and repropagation of patent-protected seed. This contrasts with the farmers' privilege under the UPOV plant breeders' rights system, where countries have the option to allow such a farmers' privilege for PVP-protected seed.

The International Seed Trade Federation (FIS) has stated that there is a need for a clear limitation or definition of those practices which are carried out in some countries under the name of 'farmers' privilege' (see *http://www.worldseed.org/~assinsel/pos_fis.htm*). FIS considers that it should not happen that a situation of unfair competition should arise between the participants in the seed market because of the use of farm-saved seed with subsequent commercial use of the product obtained – whether it comes from the production of seed or from the production for consumption purposes. FIS has also stated that, despite the benefits of the UPOV system for protection of plant varieties, it will be useful for companies to take advantage of patent protection also for plant varieties, in case of maintenance of the abusive application of the possibility for the farmer to use his own seed without paying royalties, which is considered by FIS not to be justified. This would happen because the patent system, with its stronger degree of monopolization, legally excludes such a possibility.

Nonetheless, for both social equity and food security reasons, there are justifications for providing a 'farmers' privilege' for smallholder and resource-poor farmers, especially in developing countries. This would essentially require a disaggregated 'farmers' privilege' in both PVP and patent legislation whereby poorer farmers who do not represent an immediate or lucrative market would enjoy the 'farmer privilege' to save seed, while their richer counterparts would be required to pay royalties on saved proprietary seed. Both the EU's Directive on Protection of Biotechnological Inventions (Article 11) and the Andean Pact countries have opted to enshrine the farmers' privilege for a segment of their farmers, namely subsistence or smallholder farmers, whose livelihoods can be dependent on farm-saved seed and planting materials.

PROPRIETARY RIGHTS AND THE DIFFUSION OF BENEFITS IN AGRICULTURAL BIOTECHNOLOGY RESEARCH AND DEVELOPMENT

Intellectual property right systems provide commercial incentives for increasing the rate of innovation, by allowing limited period monopoly rights over the embodied innovation. Because of the need to recoup R&D costs which are privately financed and to generate a short-term profit, there is a fundamental trade-off built within IPR systems between the rate of innovation and the rate of diffusion of the innovation to poorer clients (Barton and Strauss, 2000). Low cost and rapid diffusion of a desirable technology can be forgone for the sake of attracting increased investments into further innovation. The costs of acquiring the innovation during its early distribution phase are usually higher. This section surveys the possible impact of increasing proprietary rights on the rates of diffusion of useful technologies to poorer farmers. It also suggests some mechanisms for avoiding unnecessary restraints on technological diffusion, as they currently emerge within the commercial and IPR systems.

IPRs and the Agricultural Biotechnology Innovation System

Most innovations in biotechnology are developed using the knowledge or technologies generated from previous innovations (Scotchmer, 1991). Many plant biotechnology products or techniques are 'modular', in that they are assembled from a number of previously developed technologies/transgenes, each of which may be subject to a separate patent. The commercialization of many proprietary biotechnology products is typically contingent on other proprietary biotechnology products or processes, and in particular on agreements between IPR holders regarding the relative contributions of different proprietary technologies to the product in question. Many biotechnology products (such as transgenic seeds or transgene cassettes) now have a complex IPR pedigree because a large number of proprietary products or processes are involved in developing the product.

The commercialization of many plant biotechnology products will be dependent upon proprietary technologies owned by third parties. Therefore both companies and public sector research institutions involved in plant biotechnology research will increasingly use their own patented technologies as bargaining or trading chips for access to other useful proprietary technologies. Most research institutions in OECD countries now have patent lawyers or specialized technology transfer units who negotiate terms of access to technologies or germplasm developed or acquired by the institutions. This applies to institutions both in the private sector (such as companies) and in the public sector (such as universities).

Although knowledge is growing, the extent to which public sector agricultural biotechnology research institutes are now working with proprietary materials or technologies can sometimes be unknown from a legal standpoint. In some instances, public sector researchers may be unaware (until the point of commercialization) that some of the products or processes they are working with are patented. For instance, there is currently no efficient process for the genetic transformation of crop plants (that is, to make transgenic crops) which is not patented. Some organizations may be unknowingly conducting research using patented technology which is not under licence. Whether many public sector researchers are now working with 'unexploded' patents which will become apparent upon widespread commercialization will largely depend on the propensity and financial ability of the patent holders to enforce their patents. Of particular concern to all plant biotechnologists is the 'freedom to operate', which can be loosely defined as legal access to all the technologies required to launch or commercialize a product.

Patents and Access to Key Enabling Technologies/Research Tools

Broad monopoly rights on key or early innovations which are unduly restrictive can stifle later innovation (Dam, 1994; Jorde and Teece, 1992; Merges and Nelson, 1990). In particular, it is thought that both overbroad patents and patents covering basic research tools (enabling technologies) may discourage incremental and follow-up research (Barton, 1998a, 1998b). In this respect, the technologies used to develop biotechnology-derived products can be broadly divided into two major groups: genes and 'enabling' technologies or research tools. The genes or combinations of genes are typically responsible for the agronomic trait, whereas the 'enabling technologies' are highly useful research tools which are routinely used for the actual research and development process, irrespective of what genes are being focused upon.

Modern biotechnology research in both private and public sector institutions is increasingly reliant on a wide range of capital-intensive research tools and processes. Among these technologies there are key 'enabling technologies' which include plant transformation systems, selectable markers, gene expression technologies, gene silencing technologies and microarray/DNA chip technologies. Such basic research tools are highly valuable in themselves because they can increase the value and speed of research and development. Access to cutting-edge research tools can confer competitive advantage for any research group. Hence access to improved research tools is continually sought after.

While a wide range of research tools, products and processes are still in the public domain, in the past decade many useful enabling technologies have become increasingly proprietary as a result of successful patent applications. Many 'enabling' technologies or techniques for conducting plant molecular biology research are currently subject to patents. For instance, all existing plant transformation technologies used to generate transgenic plants are proprietary and under the control of a small number of companies. Any commercialization of transgenic plants by non-patent holders without appropriate royalty payments may run the risk of patent infringement. In essence, one researcher's research tool may be another researcher's end product which has a commercial and marketable value. Hence the distinction between basic and applied research regarding the development of enabling technologies can now be difficult to define.

It is precisely because of their value that many improved research tools and processes are patented. While the majority of such patents are held by private companies, some public sector institutions (such as NARs) and semi-public sector institutions (such as universities) in OECD countries also patent any enabling technologies they develop. One problem is that the holders of patents on such research tools will choose to licence them on an exclusive basis rather than on a non-exclusive basis, which could have a stifling effect on the research activities of other institutions or companies. Another risk is that patent holders will use a device employed by some biotechnology firms of offering licences that impose 'reach-through' royalties on sales of products that are developed in part through use of licenced research tools, even if the patented inventions are not themselves incorporated into the final products. So far, such patent holders have had limited commercial success with such reach-through licence agreements (RTLUs).

There have been complaints from both public and private sector researchers that the owners of rights to some research tools or 'enabling technologies' are reluctant to share them for research purposes (restriction of the research exemption). In June 1997, the US National Academy of Sciences communicated its concern about broad patents being issued for particular research tools such as DNA sequences (for example, ESTs) to the US Patent and Trademark Office (Anon, 1997). Expressed sequence tags (ESTs) are partial DNA sequences from genes which act as unique identifiers of each individual gene. In modern genomics research, thousands of different ESTs, each corresponding to a different gene, can be generated rapidly to give a snapshot of the majority of genes in a particular organism. The concern regarding EST patents was triggered in 1991 when the US National Institute of Health (NIH) filed its first patent application on ESTs. Despite the later withdrawal of the patent applications, the concern over access to DNA sequence information continued to generate debate, both in the USA and internationally. This debate resurfaced when the US Patent and Trademark Office on 6 October 1998 issued the first patent for an EST (Eisenberg, 1998). In the European Union, it is unlikely that ESTs on their own will qualify for patentability under the EU's Directive on the Legal Protection of Biotechnological Inventions (Article 5). It is also now being questioned whether EST patents are likely to generate a profit over and above the high cost of filing patents on each individual EST (Anon,1999).

There are now other examples of scientific concern regarding patents over useful enabling technologies. For instance, while DuPont originally made available its powerful cre-loxP recombination technology without licensing, it then changed to requiring all researchers to buy or negotiate a licence. However, DuPont is now providing this technology to some public research institutions for medical but not for agricultural research purposes (Anon, 1998a). There has also been some concern among scientists regarding proposed changes to IPRs regarding databases. In October 1996, the US National Academy of Sciences communicated to the US Department of Commerce its concern over the potential for proposed changes to intellectual property law over databases to have a negative impact on scientific research. In some instances, such as regarding the technologies necessary to make transgenic plants, it is not known whether the proprietary technologies which are in widespread use in plant biotechnology laboratories will be enforced by their owners upon commercialization. All of the commonly used methods for generating transgenic plants (*Agrobacterium*, gene guns and so on) are currently under patent and it is feasible that licence fees may have to be paid to their owners upon commercialization of transgenic plants (Hayenga, 1998). The European Science Foundation (ESF) has stated its concern that a draft EU directive proposing changes to European copyright laws regarding electronic publishing could weaken the 'fair use' arrangements for researchers (Butler, 1999).

In the USA, concern over these issues among the scientific community has led to the NIH establishing a Working Group on Research Tools, which concluded that current negotiations over intellectual property rights are burdensome and that current IPR trends pose a serious threat to biomedical research and development. The NIH panel urged a major review of the way patent law is applied in biotechnology in order to prevent overbroad patent claims. The panel also made a series of practical recommendations for publicly funded institutions that include (a) the use of standard MTAs and (b) the development and promotion of guidelines for recipients of public funds on what terms are reasonable in licences and MTAs, both for importing and exporting research tools (*www.nih.gov/news/researchtools/index.htm*).

In the agricultural arena, separate reviews have also been conducted by the US Land Grant universities and the CGIAR of the extent of reliance of their agricultural biotechnology research efforts on proprietary technology. A preliminary poll of CGIAR centres determined that, in almost 50 per cent of the cases where IARC centres were using proprietary biotechnologies, there was some uncertainty regarding whether the results of their research could be applied freely without patent infringement (Cohen *et al.*, 1998; TAC, 1998).

Preferential Access to Proprietary Agricultural Biotechnologies?

The cost of licensing patented biotechnologies can vary widely and is a factor in determining whether plant biotechnologies will represent a cost benefit, over and above conventional plant breeding approaches. In some instances key proprietary biotechnology tools have been non-exclusively licenced with very low fees. This has been the case with the Cohen-Boyer patent on recombinant DNA technology (Horton, 1998). However, the cost of licensing can often vary according to the intended use of the patented technology and some companies or patent owners will licence their technologies along a continuum, as follows:

- licences granted only in exchange for access to other useful proprietary technologies (that is, joint-licensing),
- full royalty licences (for example, for any organization or competitor who can pay),

- limited royalty licences (such as for public sector/private sector partnerships),
- royalty-free licences (such as for non-commercial crops or non-commercial markets).

Somewhat paradoxically, it would appear that research which is specifically aimed at the poorest sectors or markets in society may have access to proprietary technologies in the most preferential manner, either freely or on concessional terms. There are reasons why owners might benefit from licensing their technologies, even at no cost (demonstration of the technology, creation of demand, provoking introduction of regulations, development of partnerships). Most patent holders have an obvious interest in achieving wider demonstration of the applicability and advantages of the patented technology, even in countries/sectors where significant licence revenues might be unlikely, especially in the early years of establishing acceptability of a technology which has been controversial.

Commercially oriented IPR owners are unlikely to licence in situations where losing control of the technology damages them, either technically or financially. Licensing is highly unlikely if the licenced technology is used to compete with the licensor in profitable markets (TAC, 1998). However, there are fewer economic obstacles to free or concessional licensing of technology for research and development, provided the resulting products did not undercut the licensor in markets where there was a profit to be made.

The Importance of Segmented Markets for Preferential Access

The ability to segment non-commercial markets from commercial markets will be crucial to negotiating preferential access to patented technologies (TAC, 1998). This can be equated with the disaggregation of clients to identify who the intended primary beneficiaries of the research and development focus are. Where markets or clients can be defined (for example, resource-poor farmers, orphan and underutilized crops, non-export production) which do not represent a commercial threat to the market of the company holding the patented technology, lower to no-cost licensing of patented technologies may be possible. Such market segmentation approaches could feasibly increase access to proprietary technologies for poorer sectors of society. However, it will be difficult to negotiate such access if there is any possibility of proprietary technology leakage into potential commercially lucrative markets. Economic analysis of the international trade in crop commodities suggests that research exemptions might be easier to obtain for research to improve domestically consumed crops than for internationally traded export crops. In theory, it should be easier for public sector researchers to gain access to proprietary technologies for improvement of non-exported orphan crops such as plantain, cassava, yams or cowpea than internationally traded crops such as maize, wheat and rice (Binenbaum and Wright, 1998).

Segmentation of target crops into export and non-export crops may therefore have a major bearing on access to proprietary tools and technologies, with access being easier for non-export crops which remain below a threshold value of competition with the technology holders (ibid.). The same probably applies to varieties which exhibit broad or specific adaptation, with the latter representing less of a threat to the patent holders of 'technology leakage'. Similarly, research specifically aimed at farmers or rural groups which express little to no effective demand in financial terms is likely to have easier access to proprietary technologies than research for richer commercial farmers which may represent a lucrative target market for companies.

Market segmentation requires indicators that can clearly differentiate between commercial and non-commercial markets (or domestic and export markets). Such indicators could include the income of the farmer (Kryder *et al.*, 2000), the size of the farm holding, the crop (Wambugu, 1996) or variety (Thaim, 1998), amongst others (Byerlee and Fischer, 2000). Research organizations wishing to meet the needs of a poorer clientele will need to identify more precisely the ultimate beneficiaries of their research, if they are to use market segmentation as a tool to negotiate access to proprietary technologies. Because markets change over time, it is likely that any access to proprietary technologies, based on a market segmentation argument, is likely to be time limited. Because patents are generally valid for a period of 20 years, the time limitation would be likely to be less than that.

If markets are not segmented, proprietary technology holders can supply their technologies at only one price, which will be more than the poorest can afford. In this respect regulations regarding parallel imports are of importance. Parallel imports refers to cross-border trade in a product without the permission of the manufacturer or publisher. Such parallel imports can take place when there exist significant price differences for the same good in different markets. For instance, there are substantial price differences for the same pharmaceutical products in different countries. The advent of electronic commerce over the Internet is likely to increase the potential for parallel importation practices. In general, parallel imports are permitted under current international agreements on intellectual property under the so-called 'exhaustion' or 'first sale' doctrine, whereby the owner of intellectual property cannot control the resale of a legally purchased good, and hence parallel imports are legal. Under Article 6 of the WTO TRIPs agreement, countries are permitted to decide for themselves how to handle the issue of parallel imports. Regional trade agreements such as NAFTA or the European Union have their own rules regarding parallel imports. If proposals for 'universal exhaustion' of patent rights under the revised TRIPs agreement are agreed, it will be more difficult to segment markets and allow differential terms of access to proprietary technologies (TAC, 1998).

The International Service for the Acquisition of Agri-Biotech Applications (ISAAA) has been instrumental in facilitating technology transfer negotiations whereby proprietary biotechnologies have been made available to non-commercial or otherwise segmented markets (see *http://www.isaaa.org*). ISAAA-brokered examples include the provision of Monsanto's potato virus resistance technology for incorporation by CINVESTAV into some virus-susceptible Mexican potato varieties. Most of the ten varieties (such as Alpha) which could be transformed with the virus resistance transgene under the agreement were varieties which are not grown in moderate climates and export of transformed potatoes to the USA was reported to be explicitly excluded from the licence agreement, as was any transformation of processing varieties other than Alpha (Commandeur, 1996). In essence, the proprietary technology was made available for domestic production purposes only, including some varieties (Rosita and Nortena) that were popular among small-scale farmers in Mexico. ISAAA has also facilitated the transfer of Monsanto's transgenic sweet potato virus resistance technology to Kenya on a royalty-free basis.

There are a number of examples of such transfer of proprietary technologies to non-commercial markets in the medical sector. The World Health Organisation (WHO) has entered into partnerships regarding consessional use of proprietary drugs with SmithKline Beecham (to use albendazole against elephantiasis) and Novartis (to donate necessary medication to cure all leprosy cases detected in the next six years). Merck has committed itself to an open-ended donation programme for preferential licensing of the use of ivermectin to treat river blindness in West Africa. Hoechst has signed a partnership with WHO and Médecin Sans Frontières to make over the patent rights to the drug eflornithine, which can treat African sleeping sickness (trypanosomiasis).

Compulsory Licensing

Patent systems use rules of law that attempt the difficult task of distinguishing between inventions that would occur even without patents and inventions that require the incentive of a patent. These legal rules call for a comparison between the invention and the 'prior art', or pre-existing knowledge in the field. In some instances, the granting of a partial monopoly right to a patent holder can raise concerns about the supply of the market at reasonable prices. In response to those concerns, most national patent laws provide for some overriding conditions called 'compulsory licences'. Compulsory licences are granted by a government for the use of particular patents, copyrighted works or other types of intellectual property for particular purposes. Such compulsory licensing can be used as an instrument to promote competition in antitrust situations. However, studies have shown that some companies that are subjected to compulsory legislation change their IPR strategy to one of relying on trade secrets to protect their inventions.

The Paris Convention for the Protection of Industrial Property states that each country shall have the right to take legislative measures providing for the grant of compulsory licences to prevent the abuses which might result from the exercise of the exclusive rights conferred by the patent; for example, failure to work the patent. For instance, Canada currently allows for the granting of such licences on several grounds, including the failure to work on a commercial scale in Canada (Patent Act P-4, Article 66). The WTO's TRIPS agreement also provides for compulsory licensing of patents, but imposes some restrictions regarding the circumstances under which compulsory licensing may be applied (Article 31). These include circumstances whereby the patent is considered to be practised in an anti-competitive manner or where the patent would be practised only for the supply of the domestic market.

The EU's Directive on Protection of Biotechnological Inventions (Article 12) also provides for compulsory cross-licensing in situations where a breeder cannot acquire or exploit a plant variety right without infringing a prior patent. In such instances, the breeder may apply for a compulsory licence for non-exclusive use of the patent. If, to exploit the variety, the breeder needs a licence from the patent holder but has been refused one, a compulsory licence must be granted, 'subject to payment of an appropriate royalty'. In the EU Directive a symmetrical compulsory licensing provision also applies in situations where a patent holder cannot exploit their invention without infringing a plant variety right. In such instances, the patent holder can apply for a compulsory licence for non-exclusive use of the protected plant variety. These provisions are dependent on the proviso that the applicants have applied unsuccessfully to the patent/PVP holder to obtain a contractual licence and that the new variety or invention constitutes significant technical progress of considerable economic interest.

If current IPR trends continue in the agricultural biotechnology sector, it is possible that compulsory licensing may be invoked by some governments to promote broader access to key proprietary 'enabling' biotechnologies which, if restricted, would have a negative effect on innovation, competition and/or the 'public good'. As compulsory licensing would be a national legislative issue, it would probably require the national definition of 'public good'-type criteria under which compulsory licensing would be necessary. Each country will have its own priorities for compulsory licensing. In the USA and Europe, there is much interest in compulsory licensing for broad biotechnology patents, research tools and enabling technologies and dependent patents, and as a potential remedy for unreasonable prices. In some developing countries there is much interest in the use of compulsory licensing to obtain lower prices for pharmaceuticals (for example, for AIDS and tropical illnesses), vaccines and other essential medicines (see *http://www.cptech.org/ip/health/cl/*).

Some agreement was reached at the December 1999 WTO meeting in Seattle regarding increasing the access of the poorest countries to essential proprietary medicines, including through the use of compulsory licensing (MSF *et al.*, 1999). It is likely that continuing disputes concerning compulsory licensing will eventually come before the WTO's dispute resolution framework. In the broader IPR context, there are also continuing bilateral disputes between regions and countries regarding national differences in the interpretation and enforcement of intellectual property protection. For instance, the US government maintains a Special 301 Priority Watch List of countries where it is considered that there is a lack of adequate and effective intellectual property protection (see *http:www.ustr.gov/releases/1998/05/98-44.pdf*).

However, it is likely that most compulsory licensing processes will end up in litigation over the adequacy of remuneration, among other issues. The TRIPS agreement requires that both the decision to licence compulsorily, and the setting of remuneration provided by the licence be open to judicial review (Article 31 (*i*) & (*j*)). Governments considering compulsory licensing would have to be prepared for judicial reviews in their domestic courts, and possibly panel hearings at the WTO, where the burden of proof will be on them to justify the adequacy of remuneration.

PROVIDING INCENTIVES FOR THE DIFFUSION OF BIOTECHNOLOGY'S BENEFITS

The preceding sections indicate that the diffusion of the benefits of biotechnology constitutes an important problem for poorer societal groups. Without 'win–win' trade-offs between what is both socially and commercially desirable, increasing rates of innovation combined with increasing amounts of proprietary control may result in the disenfranchisement of a substantial portion of the world's poor, because they would not be allowed to participate in the gains from the technological advances affecting their richer counterparts. How is it possible to address the potential problems associated with increasing proprietary control over agricultural technology? This section surveys several possible methods and approaches for addressing this problem.

Public Domain Plant Biotechnologies for Poorer Farmers?

In the past, publicly funded plant biotechnology research has often used prior publication in scientific journals to ensure that publicly funded research and technologies are placed in the public domain. This approach has resulted in the availability of some useful and quite functional agricultural biotechnologies which are not patented and hence often freely accessible with minimum conditionalities. The details of many of these can be found in the extensive scientific literature where, if they are published prior to a patent application, this can render the technologies unpatentable.

Many public sector plant biotechnology groups have adopted the prior publication approach to ensuring that technologies remain in the public domain. For instance, a key biotechnology laboratory at ETH in Zurich, Switzerland, which was performing research of relevance to developing countries, adopted this approach. Similarly, the Cassava Biotechnology Network (CBN) IPR policy is based on a preference for publication and early disclosure rather than on IPR protection. Indeed, Internet publishing of data from some public sector genomics research is increasingly used to ensure that data are placed in the public domain. As a result, public sector biotechnology research is generating large amounts of publicly available data and information. Hundreds of publicly accessible biological databases now exist on the Internet, although there are indications that some public databases are now on the verge of financial collapse owing to lack of public sector funding (Ellis and Kalumbi, 1998). However, such prior publication approaches no longer ensure that the published research or technology in its original or a derived form will remain in the public domain. As a result, a number of additional approaches have been used or proposed for those situations where scientists wish legally to ensure more open (or closed) access to publicly funded research and technologies.

Material transfer agreements (MTA) are forms of contracts that typically delimit what can be done with exchanged genetic material or technologies. MTAs are a contract between two or more parties whose principal clauses are often (a) restricted sharing with third parties and (b) mandating an agreement be reached if the shared materials are subsequently to be commercialized or used for specific purposes. MTAs are legal contracts and may be used for non-patented or patented materials. MTAs or technology transfer agreements (TTAs) are now routinely used in transactions of research materials or data between biotechnology research institutions in both the public and the private sector. Depending on the objectives of the research institution's technology transfer unit, it may or may not be possible to develop derived proprietary products from the transferred material or technologies. In many instances the limits of what can and cannot be done with the transferred research materials or technologies is legally specified in considerable detail.

To prevent exclusive appropriation of publicly funded research, it may be possible to require the use of a standard MTA for the transfer of any research materials or technologies immediately resulting from such funding. Another alternative to ensure that the benefits of public sector funding are widely accessible may be to specify that any patented products or technologies directly resulting from the publicly funded research will have certain types of exemptions. For instance, this approach is used in the IPR policy of the Rockefeller Foundation, which states that IPR-protected materials and technology resulting from Rockefeller Foundation supported research will be available at zero royalty rates for use in developing countries. Similarly one research centre, ILTAB, has a policy of free access by cassava-growing developing countries to relevant proprietary technology it develops regarding cassava improvement. To ensure broader access to proprietary biotechnologies developed using public funding, publicly funded institutions could, for instance, be prevented from granting exclusive rights to any technologies which are of importance to national or global food security.

Another MTA-type approach to promote access to technologies that is being used in the computer software industry is called 'copylefting' and is typically promoted by an organization called the Free Software Foundation (FSF). The simplest way to make a software program free is to put it in the public domain, uncopyrighted. This allows people to share the program and their improvements freely, if they wish to. But it is still possible for people to convert the program into proprietary software, through making changes, many or few, and distributing the result as a proprietary product. In essence, people who receive the program in that modified form do not have the freedom that the original author gave them by placing the software in the public domain. The FSF aims to give all users the freedom to redistribute and change GNU/Linux systems software. However, instead of simply putting GNU software in the public domain, they 'copyleft' it.

Copylefting requires that anyone who redistributes the software, with or without changes, must pass along the freedom to further copy and change it. To copyleft a program, first it is copyrighted; then distribution terms are added, which are a legal instrument that gives everyone the rights to use, modify and redistribute the program's code or any program derived from it, but only if the distribution terms are unchanged. In this manner, the code and the freedoms become legally inseparable. There have been suggestions that copylefting might be used in some cases to promote greater research freedom in agricultural biotechnology, especially regarding enabling technologies (Jefferson, 1994). Such MTA-type approaches for publicly funded research represent the other end of the technology 'accessibility' spectrum compared to trade secrets or exclusive licensing of patents.

Some public sector agricultural research institutions have become involved in 'defensive' patenting of technologies they develop which may have commercial value, or might have some value in the future. For instance, the Cassava Biotechnology Network recognizes that IPRs can offer protection against misappropriation of technologies developed within the Network. CIMMYT has patented its research on apomixis to help ensure access of farmers in developing countries to such apomictic technologies (Hawtin and Reeves, 1998).

In the absence of standardized MTAs for what can be done with research materials and technologies developed from publicly funded research, it is likely that defensive patenting on an institution-by-institution basis will become a common feature for all of those agricultural research institutions (such as NARs) that can afford the legal costs of filing and defending patents (Byerlee and Fischer, 2000). Such costs may not be trivial. The CGIAR is establishing a biotechnology transfer unit with expertise in IPR law in an effort to strengthen its negotiating position with other IPR holders of useful biotechnologies. Defensive patents may also have value as bargaining chips to gain preferential access to other proprietary technologies from other institutions or companies (Barton, 1998a, 1998b). A similar type of approach is pursued by the Centre for the Application of Molecular Biology to International Agriculture (CAMBIA), which makes its proprietary biotechnologies freely available to public sector scientists in developing countries but charges private sector scientists a licensing fee.

While there are still non-proprietary biotechnologies available, the cost of access to patented technologies is likely to be a growing issue for many public sector research institutions (Byerlee and Fischer, 2000). It is illustrative that commercially oriented research in many biotechnology companies now has to follow the research route of least cost in terms of royalty payments to other companies for enabling technologies used to develop a commercial product (Mascarenhas, 1998). Unfortunately, no public sector body has yet compiled a directory of those useful plant biotechnologies which are freely accessible in the public domain, especially for scientists in developing countries.

Conversely, there is a corresponding lack of publicly available studies on what the current patent situation is for key enabling biotechnologies. However, in 1998, the CGIAR Panel on Proprietary Science and Technology conducted a study of proprietary science and technology within the CGIAR system (CGIAR, 1998). This CGIAR study included an initial review by ISNAR of the extent of use of proprietary plant biotechnology tools in each of the IARCs (Cohen *et al.*, 1998). The US Land Grant universities are conducting a review of what proprietary plant biotechnology tools are used within the Land Grant universities and under what terms. The International Society for Plant Molecular Biologists (ISPMB) is currently undertaking a comprehensive study of different patent 'families' regarding key areas of plant biotechnology in order to provide publicly better information regarding the current patent situation for its members.

The incredible escalation of MTAs in the research sector would suggest that MTAs (and the downstream transaction costs of the legal paper chase later) will be in widespread use well into the future. Harmonization of MTAs within the public sector would considerably reduce the transaction costs that will be associated with MTAs in the future. Such a process of harmonization may also lead to a coalition of interests within the public sector and define a more common front for negotiating public sector access to proprietary technologies from the commercial sector. There may be a role for organizations such as IBS, CAMBIA in such initiatives which could be either national, international or sectoral coalitions.

In the agricultural biotechnology arena, IPRs will have an increasing influence on any institutions' access to the proprietary technologies of others. Institutions with large portfolios of relevant IP will be in a better negotiating position to gain access to the IP of others. As IPRs are in greater use by the private sector than by the public sector, it is likely that public sector research institutions such as the CGIAR, NARs and individual university researchers will not be in a strong negotiating position regarding access to useful proprietary plant biotechnologies.

Will a Generics Industry Emerge in the Agricultural Biotechnology Sector?

Over time, proprietary agricultural biotechnologies are likely to exhibit analogous market trends, and product life cycle dynamics, to those observed in the pharmaceutical sector (Correa, 1997; Reddy, 1997; CBO, 1998). The agricultural biotechnology sector is at an early stage of product development where the majority of products are still under patent. As a result, a generics industry has not yet emerged in the agricultural biotechnology sector. It may transpire that the benefits of proprietary agricultural biotechnologies will only reach poorer farmers once the costs of R&D have been recouped by the patent holders after selection of more lucrative markets.

The manufacture of 'generic pharmaceuticals' for which the original patents have expired (for example after 20 years) has reduced the costs of drug delivery to poorer clients for public health reasons. The reduced costs are possible because the generic drug manufacturers do not have to pay for the high costs of both R&D and drug registration that are necessary in the early stages of any drug development. While a 'generics' market has not yet emerged in the agricultural biotechnology sector, it will undoubtedly begin to emerge within the next decade as some of the earlier generation biotechnology tools and products will come off patent (for example, the first transgenic plants were made in 1986).

The evolution of the generic pharmaceutical drug market is therefore of interest. Prescription drugs can be divided into two categories: innovator drugs and generic drugs. Innovator drugs (also referred to as brand-name drugs) generally have a patent on their chemical formulation or on their process of manufacture. They have been approved after extensive clinical testing. While they are still under patent protection, innovator drugs are called single-source drugs, because only the company that holds the patent produces them. After the patent has expired, generic copies of the exact chemical formulation usually become available. Then such drugs are referred to as multiple-source drugs. There are a range of strategies that can be employed to extend temporarily the effective patent lifetime of a drug (Robertson, 1999).

Generic drugs are a copy of an innovator drug, containing the same active ingredients, that the regulatory authorities judge to be comparable in terms of such factors as strength, quality and therapeutic effectiveness. Generic copies may be sold after the patent on a brand-name drug has expired. Generic drugs are generally sold under their chemical name rather than under a brand name. They typically obtain regulatory approval under a shorter process than innovator drugs. They are required only to demonstrate 'bioequivalence' to an innovator drug; that is, to show that the active ingredient is released and absorbed at the same rate for the generic drug as for the corresponding innovator drug. Because they are copies rather than original formulations, generic drugs are not patentable.

Competition in the pharmaceutical market currently takes three forms: among brand-name drugs that are therapeutically similar, between brand-name drugs and generic substitutes, and among generic versions of the same drug. Manufacturers of brand-name drugs compete for market share primarily through advertising and the quality of their products (including efficacy and side-effects) as well as through pricing. Manufacturers of generic drugs increase their market share mainly by lowering prices. In general, companies produce either generic or brand-name drugs, not both, although some generic manufacturers are subsidiaries of brand-name manufacturers.

Patents do not grant complete monopoly power in the pharmaceutical industry. The reason is that companies can frequently discover and patent several different drugs that use the same basic mechanism to treat an illness. The first drug using the new mechanism to treat that illness – the breakthrough drug – usually has between one and six years on the market before a therapeutically similar patented drug (sometimes called a 'me-too' drug) is introduced.

Generic medications began to be developed in industrialized countries in the 1970s as the most profitable patented medications were released into the public domain and manufacturers of generics began a price war amongst themselves and against the original developers whose drugs were now in the public domain. In countries like the USA, where there are strong incentives to replace patented medications with generics (high prices for patented medications and laws favouring competition) as soon as their patents expire, generic drugs currently make up half the pharmaceuticals market.

Worldwide, the pharmaceutical market has become increasingly competitive since the early 1980s, in part because of the dramatic growth of the generic drug industry. In 1996, 43 per cent of the prescription drugs sold in the USA (as measured in total countable units, such as tablets and capsules) were generic. Twelve years earlier, the figure was just 19 per cent. Generic drugs typically cost less than their brand-name, or 'innovator', counterparts. Thus they have played an important role in holding down national spending on prescription drugs from what it would otherwise have been. The US Congressional Budget Office (CBO) estimates that by substituting generic for brand-name drugs, purchasers saved roughly $8 billion to $10 billion in 1994 (at retail prices).

Greater sales of generic drugs reduce the returns that pharmaceutical companies earn from developing brand-name drugs. The 1984 Hatch–Waxman Act in the USA aimed to limit that effect by extending the length of time that a new drug is under patent, and thus protected from generic competitors. Those extensions compensate for the fact that for part of the time a drug is under patent it is being reviewed by the Food and Drug Administration (FDA) rather than being sold. The act tried to balance two competing objectives: encouraging competition from generic drugs while maintaining the incentive to invest in developing innovative drugs. The Hatch–Waxman Act also eliminated the duplicative tests that had been required for a generic drug to obtain approval from the FDA. The Hatch-Waxman Act reduced the average delay between patent expiration and generic entry from more than three years to less than three months for top-selling drugs.

Manufacturers of prescription drugs can be divided along similar lines: companies that primarily produce innovator drugs and companies that focus on generic drugs. The two types of manufacturers compete very differently. Producers of innovator drugs invest heavily in research and development (R&D), hoping to recoup that investment in profits from future sales while a drug is under patent and they have a monopoly on its manufacture. Producers of generic drugs do not need to duplicate the research effort of the innovator firm or to invest nearly as much in getting regulatory approval for their drugs. Such cost savings are substantial because developing an innovative drug is a major industrial undertaking, with major investments of time (often a decade or more) and of capital (between $16 million and $500 million) required.

Although companies invest in research and development because they expect high returns from the future sales of their discoveries, those returns are considerably skewed. Some drugs have billion dollar sales, whereas others bring in less than $25 million a year. For drug manufacturers to be successful, the present value of their future profits from the sale of new products (discounted to the date the products were introduced) must exceed the capitalized cost of their original R&D investment (capitalized to the date of market introduction), including investment in drugs that never make it to the market. Patents increase the rewards for innovation by giving companies a temporary monopoly over marketing their discoveries. That temporary monopoly status is often necessary to provide sufficient incentives for drug companies to invent the new products that benefit consumers. Without patents, many new drugs could be easily and quickly duplicated by other manufacturers, preventing the innovator firm from obtaining enough reward to justify its investment.

Since generic producers have neither patents nor a costly approval process to deter potential competitors, they quickly face competition from other companies producing identical drugs. That intense competition forces generic manufacturers to charge much lower prices than the innovator firm, which (even after its patent expires) typically enjoys a market advantage based on its reputation for producing a high-quality product. Because of the cost factors involved, generic drug manufacture is considered to be a major factor in ensuring broader access to pharmaceuticals among poorer social groups in developing countries. Many developing countries have created national pharmaceutical industries to replace costly imports and to supply the country's needs at the lowest prices for social and public health reasons. Some countries, such as India and Egypt, have preferred to support locally financed drug manufacturing enterprises and, like most other countries, have expressly excluded medications from their national patent legislation (South Korea being among the others). Within the next decade it is likely that generics-based strategies for agricultural biotechnology product development and diffusion to lower-income groups will begin to emerge.

Public Investments in the Diffusion of Biotechnology's Benefits

Farmers interface with the products of biotechnology or agricultural research through a range of intermediary service providers, usually through public extension or private marketing agents. Agricultural extension agents are the public sector equivalent of agricultural marketing or sales agents in the private sector. The distribution channels by which products reach farmers' fields are now undergoing major structural changes worldwide. There are now a wide range of public, private and non-governmental organizations with differing objectives attempting to deliver appropriate products to different groups of farmers. This has significant implications regarding the nature of the technology-disseminating organizations that agricultural biotechnologists interact with in identifying what priority technologies are needed, what farmers are the resultant client group and what types of farmers and consumers will ultimately reap the benefits of agricultural biotechnology research and development.

Agricultural extension is now in a process of reform and transition worldwide (Rivera, 1996). Pressures towards cost recovery and privatization have led to rapid slimming of public sector extension services in Europe, the USA and Australasia over the last decade (Rivera and Gustafson, 1991). In parallel, public sector agricultural extension services in developing countries are achieving only limited impact but face unsustainably high recurrent costs (ibid.). In many countries, governments are withdrawing national agricultural research systems (NARs) from extension services and now expect other institutions (private or non-governmental organizations – NGOs – such as farmers' organizations) to provide and/or finance such activities (Farrington *et al.*, 1993). Financial pressures have led to a search for ways of reducing public sector costs by, for example, privatizing parts of the extension service, having farmers pay the government for some services, and cost-sharing arrangements between government and NGOs such as farmers' organizations. The most efficient public sector extension services of the future are likely to focus on spheres (geographic, thematic, social) inadequately serviced by the private commercial sector (Moris, 1991a). As a result, novel extension approaches are emerging which are farmer-participatory, institutionally pluralistic and geared towards cost sharing (ibid.). For example, a range of approaches for farmer-led approaches to agricultural research and extension are now emerging which have some potential for rural poverty alleviation (Scarborough, 1996a).

It is unclear how agricultural biotechnology research could better interface with such changes, especially in relation to NGO or farmer-led approaches to agricultural research and extension. It cannot be assumed that even useful agricultural biotechnologies which are wholly and unrestrictedly in the public domain will actually reach the fields of poorer farmers in the short term through existing state extension channels. The Gatsby Charitable Foundation has recognized this in developing a 'research-managed extension' (RME) model for more effective transfer of agricultural biotechnologies to poorer farmers in developing countries. The RME model relies on a reward system based upon the intensity of contact between extensionists and farmers. Also the Netherlands Ministry of Foreign Affairs has a Special Programme on Biotechnology and Development Cooperation which has been exploring pilot projects on 'appropriate biotechnologies' which might better meet the needs of small-scale farmers in developing countries. A key feature of such approaches has been farmer participatory needs assessments to determine research priorities prior to initiation of research and development (Bunders, 1990; Bunders and Broerse, 1991). However, most plant biotechnology research is conducted far 'upstream' of such 'downstream' structural changes in the agricultural extension and marketing sectors.

There has been a lack of biotechnology research which would enable key agricultural 'processes' at the 'on-farm' level in order to improve or 'empower' poorer farmers' livelihoods. Yet, in theory at least, plant biotechnology research could be applied towards such goals, especially if there were better linkages between farmer participatory researchers/extensionists and plant biotechnologists (Thro and Spillane, 2000). The CGIAR's Systemwide Programme on Participatory Research and Gender Analysis (SWP-PRGA) is currently exploring whether some biotechnologies might have utility in 'empowerment'-oriented farmer participatory plant breeding. The Centre for the Application of Molecular Biology to International Agriculture (CAMBIA) in Australia has for some years been trying to develop plant biotechnology tools which could empower low-technology approaches to crop improvement in developing countries (Jefferson, 1993).

Technology Transfer to Developing Countries

Agricultural biotechnology research is considered to be most powerful when it is fully integrated with conventional breeding or agricultural improvement approaches (Thro and Spillane, 2000). However, agricultural research and development (R&D) *sensu latu* is undergoing major dynamic changes both in the private and in the public sector, which the agricultural biotechnology community will have to take into account in determining which social groups are likely to benefit most from an agricultural biotechnology 'revolution'.

Agricultural R&D is considered to be essential to the competitiveness of the agriculture sector in most countries and is known to generate high rates of return. Most studies of the private and social rates of return to agricultural research have concluded that they have been very high – typically more than 20 per cent per year – compared to 3–5 per cent per year for the long-term real rate at which governments borrow money. Recognizing this, the World Bank recommends that each country invest at least 2 per cent of its agricultural GDP in agricultural research and development. Most countries fail to reach this level. In general, the mix of private and public sector-funded agricultural research in any country reflects the general type of political economy promoted by the government.

However, public support for national and international agricultural research is now declining in most countries. In the past two decades there has been a decline in public funding for agricultural research, including plant and animal breeding. Towards the end of the 1980s, the growth in agricultural research and development slowed considerably, particularly in developing countries. For instance, in sub-Saharan Africa, real spending per scientist has fallen by 2.6 per cent a year since 1961, with the rate of decline accelerating from 1.6 per cent a year during the 1960s to 3.5 per cent a year during the 1980s (Anderson *et al.*, 1994). As a result, some economists believe that there is significant underinvestment in agricultural research, particularly in situations where the private sector has not become involved in research areas from which the public sector has divested its interest. Since public funds will always be scarce, especially in developing countries, those funds that are available for agricultural research and development should ideally be allocated to uses with the highest social returns.

Some developing countries have had considerable success in establishing significant capacity in biotechnologies such as plant tissue culture, micropropagation and disease diagnostics, and in meeting farmers' needs with such technologies. However, the strengthening of capacity in the plant molecular biotechnology research has proved more difficult to achieve in the short term, especially in a manner which is aimed at meeting country-specific needs. In most instances, the existence of a conventional plant breeding programme which is operational is a necessary prerequisite for any rational application of advanced plant molecular biotechnology techniques such as marker-assisted selection or transgenesis.

Despite some successes, there is a growing consensus that many international project-based initiatives to transfer biotechnology capacity to developing countries have not been as successful as originally envisaged. The majority of developing counties have limited practical access to the tools and germplasm necessary to apply high-technology biotechnology research to their national needs. The barriers to such access are many and mainly include lack of financial, scientific and infrastructural resources.

Cross-country reviews of the state of agricultural biotechnologies in some developing countries have been performed by ISNAR-IBS and OECD. These concluded that there are major differences between countries in their agricultural biotechnology capacity which would preclude any generalizations regarding the appropriateness (or not) of some biotechnologies for developing countries as a generic group. For instance, a number of developing countries in Asia and Latin America, such as Brazil, China, India, Malaysia, Thailand, Philippines and Indonesia, have a relatively high level of plant biotechnology capacity, especially in early generation biotechnology areas such as plant micropropagation, transgenics and marker-assisted breeding (Komen and Persley, 1993). In Africa, a significant number of countries (for example, Burkina Faso, Cameroon, Côte d'Ivoire, Gabon, Ghana, Senegal, Ethiopia, Uganda, Madagascar, Malawi and Zambia) have some limited biotechnology capacity in the areas of plant tissues culture and micropropagation. In some African countries, basic infrastructure and facilities for even the simplest plant tissue culture or micropropagation are not available. However, other African countries such as Morocco, Tunisia, Nigeria, Kenya and Zimbabwe have some additional but limited capacity in plant molecular biology (Brink *et al.*, 1998). While countries such as South Africa, Nigeria and Egypt have the capacity to generate transgenic plants, mechanisms to ensure that the plants can reach the end-user (that is, the farmer) are in many instances lacking (ibid.).

The OECD's Development Centre has published an excellent study on the incentives, constraints and country experiences for integrating biotechnology in agriculture in different developing country situations (Brenner, 1996). ISNAR's Intermediary Biotechnology Service has also produced a series of research reports which provide useful frameworks for decision making regarding national biotechnology priorities, planning and policies, based on the experience selected developing countries have had to date with the integration of biotechnologies into their agricultural research systems (Komen and Persley, 1993; Cohen, 1994).

The OECD study concluded inter alia that biotechnology research has not been closely integrated with the problems and constraints confronting the agriculture sector, nor has it addressed the obstacles to widespread diffusion of useful new biotechnologies, particularly to low-income farmers. A lack of clear priorities and focus was identified. The study called for reflection on the part of developing countries, scientists, NGOs, donors and the CGIAR on the development of innovative public/private mechanisms for the transfer of 'public good' biotechnologies in developing country agriculture. The study also stressed the importance of long-term public sector funding if the benefits of agricultural biotechnology research are to be realized by the poorer strata of society.

A 1994 survey of 45 organizations involved in the transfer of agricultural biotechnology revealed that most initiatives concentrate on the few developing countries with relatively advanced scientific and technological capabilities, and that developing country scientists and administrators are not always directly involved in their planning and design (Brenner and Komen, 1994). Also, a 'brain drain' exists for many developing countries and regions (for example, China, India, Africa) whereby many of their scientists have moved to work or train in advanced biotechnology laboratories in the USA, Europe, Australia and Japan. If such scientists return to situations where there is little or no conventional plant breeding activity or infrastructure, the comparative advantage that they have learnt in plant biotechnology cannot easily be applied to the improvement of agriculture in their own countries. There are currently few financial or other incentives for such scientists to return to conduct research in their countries of origin (Gbewonyo, 1997). In the absence of public sector funding for such scientists upon return, it is likely that many such scientists will become technology adapters and/or marketing agents for imported proprietary products or germplasm developed by non-domestic companies.

In all countries, there is also a need to involve more actively end-users such as farmers' and producers' organizations in priority setting regarding the objectives of publicly funded agricultural biotechnology (Braunschweig, 2000). At the international level, the International Federation of Agricultural Producers (IFAP), an organization which represents a large proportion of the world's farmers, recently made a policy statement on 'Farmers and Biotechnology' at the 1998 World Farmers Congress (see *http://www.ifap.org/biotech.htm*) The IFAP policy statement raised issues regarding (a) the potential benefits of genetically modified organisms (GMOs) to different stakeholders, (b) concerns of different stakeholders regarding GMOs, (c) promoting freedom of farmers to operate, (d) promoting safety and accurate information, (e) increasing public sector research investment, (f) intellectual property rights, (g) addressing the needs of developing countries, and (h) maintaining biodiversity.

Targeting Local Investments in Biotechnology R&D

Giving membership-based groups which are truly representative of the needs of resource-poor farmers a publicly subsidized voice in public sector agricultural research and development is an approach that could generate a much greater poverty alleviation impact for agricultural biotechnology (Bebbington *et al.*, 1994; Spillane and Thro, 2000). Such farmer or client participation in priority setting may help to reorient agricultural biotechnology towards meeting the needs of poorer farming groups that have tended to be neglected (Haugerud and Collinson, 1990). However, a number of important issues must be addressed if client-driven research is actually to meet the needs of resource-poor rather than richer farmers (Ashby and Sperling, 1994). For instance, biotechnology research could become more effective and demand-driven if farmers' organizations could support their demand for research services by funding research activities they consider of immediate and strategic importance (Collion and Rondot, 1998). It is important that such farmers' organizations are accountable to their members for the quality of their demand pull on research services (Fox, 1992). This approach has been shown to work for wealthier farmers in the Netherlands and Zimbabwe (Roling, 1989; Biggs, 1989).

In most instances the clients of publicly funded agricultural biotechnology research are non-disaggregated and hence ill-defined. To institutionalize accountability sharing in the public sector requires that identified client groups gain more control over the budget and resource allocation to publicly funded applied research so that more resources can be apportioned to those research activities which most meet the needs of the clients. While this aspect may at first seem conflict-prone from the perspective of public sector researchers, the identification and collaboration of client groups may be one means of ensuring the survival of public sector funding agricultural research in many countries (Spillane and Thro, 2000). It will also help to generate political support for agricultural biotechnology research which has a beneficial economic or social impact.

There are some examples of farmer involvement in decision-making structures of national agricultural research institutions, although the extent of involvement of resource-poor farmer groups is unclear (Arnaiz, 1995). The following represent a selection of examples of potential relevance to client-driven plant biotechnology research.

- A number of national agricultural research organizations in Mali (IER, NARC), Senegal (ISRA, ANCAR), Burkino Faso and Guinea (IRAG) are developing partnership mechanisms with farmers' or producers' organizations in order to make their research more demand-driven and client-responsive (Collion, 1995; Collion and Rondot, 1998). In Mali, Burkina Faso and Guinea, some of the producer organizations contract specific areas of research and development from the national agricultural research institutes, which is either funded by the producers' organizations (Guinea), donors (Burkina Faso) or governments (Mali, Senegal).

- The Instituto Rio Grandese do Arroz (IRGA) in Brazil is supported by taxes on rice production of about 1 per cent. Although the scientific personnel and administration are employed by the government, their decisions must be approved by a council of farmers and the private sector. Funds obtained through the tax cover not only biophysical research but also socioeconomic studies as well as administration, maintenance of research stations and other operational costs.
- Argentina's Instituto Nacional de Tecnologia Agricola (INTA) is organized similarly. INTA's administration has representatives from both the public and the private sectors, allowing research beneficiaries to participate in the establishment of priorities. INTA receives 90 per cent of its budget from a 1.5 per cent tax on all agricultural exports. It obtains the remaining 10 per cent from special projects and by selling property rights to the private sector for varieties and technologies developed by INTA. The institute's rice programme has received substantial financial support from farmers', millers' and agronomists' associations through an organization called PROARROZ. A law has been developed that would make contributions based on rice sales mandatory (INTA, 1991).
- The Instituto Nacional de Investigacion Agraria y Agroindustrial (INIAA) in Peru has a central administration and experiment stations in the most important rice-growing areas. A financial crisis in 1992 led to INIAA handing over the administration of its experiment stations (those located in the most developed areas) to farmers' cooperatives. Since then, research has been organized and sponsored by farmers and carried out by scientists on the government payroll.
- In China, 80 per cent of farmers are smallholders who farm less than half a hectare, using traditional farming practices. In 1985, the State Science and Technology Commission was established with the goal of ensuring that 70–80 per cent of applied agricultural research was actually used by farmers. To meet this goal, each province's Academy of Agricultural Sciences was required to enter into contracts with farming and trading oganizations, where such contracts might account for up to 10 per cent of the Academy's income (as in Jiangsu province). However, payment is based strictly on results, with a refund to the client farmers/traders groups if a research project fails (Forestier-Walker, 1987).
- The Chilean Agricultural Research Institute (INIA) was privatized in 1986. In the early 1990s, it began to establish contractual arrangements with NGOs and small farmers' organizations, which in some

> cases involved farmers in joint planning activities. One contracted NGO, the Agrarian Research Group, organized farmers into village-level agricultural committees of 15–40 members, under one umbrella organization which has a budget to contract research from INIA. Such committees develop technical programmes which, inter alia, are responsible for on-farm experimentation and adaptive testing of technologies developed by INIA (Berdegue, 1990).

A major problem for farmer participation in priority setting is that many farmers' organizations (especially resource-poor groups) may lack the capacity to analyse members' constraints, aggregate and prioritize needs and articulate them (Fox, 1992; Mercoiret *et al.*, 1990). Even where such needs are prioritized, they are more likely to be related to their immediate economic or institutional situation (pricing policies, land reform, credit and so on), rather than to the identification of a specific technological or trait need (Collion and Rondot, 1998). While some wealthier farmers' organizations recruit their own technical specialists to fill this gap, usually farmers' organizations only formally approach research organizations on an ad hoc basis when constraints that may have technical solutions threaten their livelihoods (Bratton, 1985). However, training can be provided to enable farmers' organizations both to develop their own technical skills and to improve their ability to identify, aggregate and prioritize members' technology needs (Arnaiz, 1995).

CONCLUSION

Food staples typically absorb half the consumption of people below the poverty line and are their main source of nutrients. There is little doubt that, if plant biotechnology research was applied to well-defined social or economic objectives such as improving the food staples of the poor, it could benefit poorer rural and urban groups. However, in an era of privatization, there remains the valid concern that the needs of poorer farmers or nations are unlikely to be a factor which favourably steers the research objectives of biotechnology research which is wholly dependent on private investment.

To date few of the products of advanced agricultural biotechnologies have been disseminated to poorer farmers. If agricultural biotechnology is to be demand-driven, the question arises regarding what institutions will meet the needs of poorer social groups. Given the lack of commercial viability of focusing on poorer people as markets, pro-poor agricultural biotechnology research and development will have to be publicly funded. Long-term public sector investment in agricultural (biotechnology) research will therefore be essential to address the needs of poorer farmers and consumers who do not constitute a significant enough commercial market for private sector biotechnology research and development.

However, there is much scope for private–public sector collaborations for pro-poor biotechnology, especially where markets can be segmented so that the profits of the collaborating private sector partners are not unduly compromised. A rational and transparent division of objectives, comparative advantages and labour between the public and private sector will also be essential to resource-allocation strategies for pro-poor agricultural biotechnology. Novel public funding approaches based on social venture capital will be necessary if priority agricultural technologies which are appropriate to the needs of the poor are to be developed. Innovative models for such financing are emerging in the pharmaceutical sector and could equally be applied to the agriculture sector. These include 'push–pull' funding strategies, orphan or neglected research area acts and the provision of incentives for increased competition within the appropriate sectors.

At the governmental level, policy instruments are currently lacking which promote or encourage biotechnological research which could contribute to food and livelihood security in resource-poor situations, especially in developing countries. Agricultural biotechnology research which is aimed specifically at segmented non-commercial markets or clients groups such as orphan or underutilised crops, non-export crops or resource-poor farmers may have better access to proprietary biotechnologies than more commercially oriented research. Hence strategic management of intellectual property by both the public and private sectors will be necessary to establish mechanisms whereby poorer social groups can benefit from proprietary biotechnologies, without compromising the ability of the private sector to earn a return on its investments.

To promote national food security, such research exemptions for orphan or underutilised crops, non-export crops or resource-poor clients could be incorporated in national policy instruments such as laws on patents, PVP and genetic resources. At present, very few biotechnology research institutions worldwide have addressed the way in which more innovative research exemptions on their proprietary technologies might be used to promote world food or livelihood security. From a public policy perspective, Barton (1998b) recommends that developing countries would be wise to enact patent laws with broad experimental use exemptions, to issue relatively narrow patents and to establish adequate antitrust/competition policies. In addition, harmonization of MTAs within the public sector, in conjunction with active market segmentation of clients, will reduce the transaction costs associated with disseminating agricultural biotechnologies to poorer social groups.

Anti-biotechnology lobbies at a global level are currently focused on stopping biotechnology research in agriculture (especially of a transgenic variety) irrespective of whether such research has a pro-poor focus. Most such opposition emanates from food surplus OECD countries. However, the poor will suffer disproportionately from the lack of public sector agricultural biotechnology research options that will result from the emerging current funding and regulatory climates. In the developed countries, both the pro- and anti- 'biotechnology in agriculture' lobbyist groups represent quite small and specific client groups (for example, company shareholders, organic producers) in society, yet increasingly have a disproportionate control over the direction of the public sector agricultural research agenda. Organizations with broader memberships such as farmers' organizations, scientific organizations or trade unions have had less agenda-setting impact in terms of funding allocations and research objectives. We may have reached the juncture where there is a need for improved 'democratization' and 'good governance' regarding policy decisions, priority setting and funding for national and international agricultural research.

The participation of farmers in research agenda setting and decision making would also help in openly developing a culture of 'accountability' whereby the reasons for choosing a particular technological or agricultural research direction are made clearer to all stakeholders. The current controversy regarding agricultural biotechnology in food surplus countries would suggest that scientists and industry have naively assumed that the benefits of the application of biotechnologies to agriculture are evident to all stakeholders. In this context, an important point is that the poor are largely (and increasingly) silent in terms of setting agendas regarding agricultural biotechnology. Poorer social groups are particularly vulnerable to being misrepresented (regarding their needs) by intermediaries with technology prejudices or biases. Giving membership-based organizations which are truly representative of (and accountable to) poorer social groups in food deficit countries a voice in the technology and agriculture debate will be essential to democratize agricultural research planning and policy.

There is a need to shift the biotechnology debate from a sterile confrontation between commercializers and critics to actively developing policies, mechanisms and/or institutions that ensure that the poor could benefit from agricultural biotechnologies. Once stakeholders are clearly identified, the real challenge for public sector research institutions with a poverty alleviation mandate is to gear biotechnological research towards the specific problems of the poor, and this will require a research agenda in which appropriate biotechnologies are part of a broader technological approach to sustainable agriculture.

REFERENCES

Agris, C.H. (1998a), 'International patent filing', *Nature Biotechnology*, 16, 479–80.

Agris, C.H. (1998b), 'Patenting DNA sequences', *Nature Biotechnology*, 16, 877.

Agris, C.H. (1999), 'Intellectual property protection for plants', *Nature Biotechnology,* 17, 197–8.

Alston, J.M. and P.G. Pardey (1995), *Making Science Pay: The Economics of Agricultural, R&D Policy*, Washington, DC: American Enterprise Institute Press.

Anderson, J.R., P.G. Pardey and J. Roseboom (1994), 'Sustaining growth in agriculture: A quantitative review of agricultural research investments', *Agricultural Economics*, 10, 107–23.

Anon (1997), 'Dangers in EST patent law', *Plant Molecular Biology Reporter*, 15, 205–8.

Anon (1998a), 'Conditionally yours', *Nature Biotechnology*, 20, 1–3.

Anon (1998b), 'Novartis pours cash into UCB', *Nature Biotechnology*, 16, 1298.

Anon (1999), 'Patenting ESTs: is it worth it?', *Nature Genetics*, 21, 145–6.

Arnaiz, M.E.O. (1995), 'Farmers' organizations in the technology change process: An annotated bibliography', Network Paper 53, Agricultural Administration (Research and Extension), ODI, London.

Arunde, A. and A. Rose, (1998), 'Finding the substance behind the smoke: Who is using biotechnology?' *Nature Biotechnology*, 16, 596–7.

Ashby, J.A. and L. Sperling (1994), 'Institutionalizing participatory client-driven research and technology development in agriculture', Network Paper 49. Agricultural Administration (Research and Extension) ODI, London.

Ashby, J.A. and L. Sperling (1995), 'Institutionalizing participatory, client-driven research and technology development in agriculture', *Development and Change*, 26, 753–70.

Barton, J. (1998a), 'The impact of contemporary patent law on plant biotechnology research', in S.A. Eberhart, H.L. Shands, W. Collins and R.L. Lower (eds), *Intellectual Property Rights III Global Genetic Resources: Access and Property Rights,* Madison, WI: CSSA.

Barton, J. (1998b), 'International intellectual property and genetic resource issues affecting agricultural biotechnology', in C.L. Ives and B.M. Bedford (eds), *Agricultural Biotechnology in International Development*, Wallingford: CABI, pp. 273–4.

Barton, J.H. and J. Strauss (2000), 'How can the developing world protect itself from biotech patent-holders?', *Nature*, 406, 455.

Bebbington, A.J., Merrill-Sands and J. Farrington (1994), 'Farmer and community organizations in Agricultural Research and Extension: functions, impacts and questions', Network Paper 47, Agricultural Administration (Research and Extension) ODI, London.

Berdegue, S. (1990), 'NGOS and farmers' organizations in research and extension in Chile', ODI Network Paper 19, Overseas Development Institute, London.

Biggs, S. (1989), 'Resource-poor farmer participation in research: a synthesis of experiences from nine national agricultural research systems', OFCOR Comparative Study Paper 3, International Service for National Agricultural Research, The Hague.

Binenbaum, E. and B. Wright (1998), 'On the significance of South–North trade in IARC crops', in TAC (ed.), *Report of the CGIAR Panel on Proprietary Science and Technology*, Document SDR/TAC: IAC/98/7.1.

Bjornson, B. (1998), 'Capital market values of agricultural biotechnology firms: How high and why?', *AgBioForum,* 1, 69–73 (*http://www.agbioforum.missouri.edu*).
Braunschweig, T. (2000), 'Priority Setting in Agricultural Biotechnology Research', Research Report no. 16, ISNAR, The Hague.
Brenner, C. (1996), *Integrating Biotechnology in Agriculture: Incentives, Constraints and Country Experiences*, Paris: OECD.
Brenner, C. and J. Komen (1994), 'International Initiatives in Biotechnology for Developing Country Agriculture: Promises and Problems', Technical paper no. 100, June, OECD Development Centre, Paris.
Brink, J.A., B.R. Woodward and E.J. DaSilva (1998), 'Plant biotechnology: a tool for development in Africa', *Electronic Journal of Biotechnology,* 1 (3), 15 December, *http://ejb.org/*.
Bunders, J.F.G. (ed.) (1990), *Biotechnology for Small-scale Farmers in Developing Countries. Analysis and Assessment Procedures*, Amsterdam: Free University Press.
Bunders, J.F.G. and J.E.W. Broerse (eds) (1991), *Appropriate Biotechnology in Small-scale Agriculture: How to Re-orient Research and Development*, Wallingford: CAB International.
Butler, D. (1999), 'Tougher EU copyright rules come under fire', *Nature*, 379, 397.
Buttel, F.H. (1986), 'Biotechnology and agricultural research policy: Emergent issues', in K.A. Dalberg (ed.), *New Directions for Agriculture and Agricultural Research, Neglected Dimensions and Emerging Alternatives*, Totowa, NJ: Rowan and Allanheld.
Byerlee, D. and G. Alex (1998), *Strengthening National Agricultural Research Systems: Policy Issues and Good Practice*, Washington, DC: World Bank.
Byerlee, D. and K. Fischer (2000), 'Accessing modern science: Institutional and policy options for biotechnology in developing countries', paper presented at the 4th International Conference on the Economics of Agricultural Biotechnology, 24–28 August, Ravello, Italy.
Carroll, C. (1992), *Intermediary NGOs: The Supporting Link in Grassroots Development*, West Hartford: Kumarian Press.
Case, J. (1995), *Open Book Management*, New York: Harper Collins.
CBO (1998), 'How Increased Competition from Generic Drugs Has Affected Prices and Returns in the Pharmaceutical Industry', Congressional Budget Office of the USA, Health Studies and Reports, July (*http://www.cbo.gov/*).
CGIAR (1998), *Report of the CGIAR Panel on Proprietary Science and Technology*, Document SDR/TAC: CGIAR//7.1, CGIAR: Washington, DC.
Coffmann, W.R. and M.E. Smith (1991), 'Roles of public, industry and international research centre breeding programmes in developing germplasm for sustainable agriculture', in CSSA (ed.), *Plant Breeding and Sustainable Agriculture: Considerations for Objectives and Methods*, CSSA Special Publication no.18, 1–9.
Cohen, J.I. (1994), 'Biotechnology priorities, planning and policies', ISNAR Research Report 6, ISNAR, The Hague.
Cohen, J.I., C. Falconi, J. Komen and M. Blakeney (1998), 'The use of proprietary biotechnology research inputs at selected CGIAR centres', Report of an ISNAR study commissioned by the Panel on Proprietary Science and Technology of the CGIAR, ISNAR, The Hague.
Collion, M.H. (1995), 'On building a partnership in Mali between farmers and researchers', AgREN Paper 54, ODI, London.

Collion M.-H. and P. Rondot (1998), 'Partnership between agricultural services institutions and producers' organisations: Myth or reality?', Agricultural Research and Extension Network Paper no. 80.
Commandeur, P. (1996), 'Private–public cooperation in transgenic virus-resistant potatoes, Monsanto, USA – CINVESTAV, Mexico', *Biotechnology and Development Monitor*, 28, 14–19.
Cook, J.D., L.S. Emptage, F.W. Miller, S. Rauch and J.-P.M. Ruiz-Fenes (1997), 'Food biotechnology: Can you afford to be left out?', *The McKinsey Quarterly*, 3, 78–89.
Correa, C. (1997) *The Uruguay Round and Drugs*, Geneva: Organisation mondiale de la santé.
Dam, K.W. (1994), 'The economic underpinnings of patent law', *Journal of Legal Studies*, 247.
Davidson, S. (1996), 'Orphan drugs: European biotechnology waits for EC to act', *Nature Biotechnology*, 14, 419–20.
Daw, M.E. (1989), 'The contribution of farming systems approaches to sustainable agricultural development', paper presented to the Farming Systems Research/Extension Symposium at University of Arkansas, 11 October.
Eisenberg, R. (1997), 'Patenting research tools and the law', Chapter 2 in *Intellectual Property Rights and Research Tools in Molecular Biology, Summary of a Workshop Held at the National Academy of Sciences, February 15–16, 1996*, Washington, DC: National Academy Press (see *http://www.nap.edu/readingroom/books/property/*).
Eisenberg, R.S. (1998), 'Do EST patents matter?', *Trends in Genetics*, 14, 379–81.
Ellis, B.M. and D. Kalumbi (1998), 'The demise of public data on the web?', *Nature Biotechnology*, 16, 1323–4.
Farrington J. (1994), 'Public sector agricultural extension: Is there life after structural adjustment?', ODI Natural Resources Perspectives no. 2, ODI, London.
Farrington, J. (1997), 'Farmers' participation in agricultural research and extension: Lessons from the last decade', *Biotechnology and Development Monitor*, 30, 12–15.
Farrington, J., A. Bebbington, K. Wellard and D.J. Lewis (1993), *Reluctant Partners? Non-Governmental Organizations, the State and Sustainable Agricultural Development*, New York: Routledge.
Forestier-Walker, K. (1987), 'China frees farming from politics', *New Scientist*, 14 May.
Fox, J. (1992), 'Democratic rural development: Leadership accountability in regional peasant organizations', *Development and Change*, 23, 1–36.
Fox, J.L. (1998), 'Mystery plea for review of seed and biotech mergers', *Nature Biotechnology*, 16, 811.
Francis, C.A. (1986), *Multiple Cropping Systems*, New York: Macmillan.
Freiberg, B. (1997), 'The Monsanto contracts: Are they good or bad?', *Seed and Crops Digest*, 48, 3.
FTC Report Anticipating the 21st Century: Competition Policy in the New High-Tech, Global Marketplace (*http://www.ftc.ov/opp/global.htm*).
Gbewonyo, K. (1997), 'The case for commercial biotechnology in sub-Saharan Africa', *Nature Biotechnology*, 15, 325–7.
Gill, G.J. and D. Carney (1999), 'Competitive agricultural technology development funds in developing countries', ODI Natural Resource Perspectives no. 41, ODI, London.

Goldberg, R. (1999), 'The business of agriceuticals', *Nature Biotechnology*, Supplement 17: BV5–6.

Gonsalves, D. (1998), 'Control of Papaya Ringspot Virus in Papaya: A Case Study', *Annual Review of Phytopathology*, 36, 415–37.

Goto, F., T. Yoshihara, N. Shigemoto, S. Toki and F. Takaiwa (1999), 'Iron fortification of rice seed by the soybean ferritin gene', *Nature Biotechnology*, 17, 282–6.

Gressel, J., J.K. Ransom and E.A. Hassan (1996), 'Biotech-derived herbicide resistant crops for third world needs', *Annals of the New York Academy of Sciences*, 792, 140.

Haugerud, A. and M. Collinson (1990), 'Plants, genes and people: Improving the relevance of plant breeding in Africa', *Experimental Agriculture*, 26, 341–62.

Hawtin, G. and T. Reeves (1998), 'Intellectual property rights and access to genetic resources in the Consultative Group on International Agricultural Research', in S.A. Eberhart, H.L. Shands, W. Collins and R.L. Lower (eds), *Intellectual Property Rights III: Global Genetic Resources: Access and Property Rights*, Madison, WI: CSSA.

Hayenga, M. (1998), 'Structural change in the biotech seed and chemical industrial complex', *AgBioForum*, 1, 43–55 (see: *http://www.agbioforum.missouri.edu*).

Hodgson, J. (2000), 'Moratorium hits Danish companies', *Nature Biotechnology*, 18, 139–40.

Horton, B. (1998), 'Taking knowledge from bench to bank', *Nature Biotechnology*, 395, 409–10.

INTA (1991), 'INTA 35 anos de tecnologia para el agro argentino', Buenos Aires.

James, C. (1997), 'Global status of transgenic crops in 1997', ISAAA Briefs no. 5, ISAAA, Ithaca, NY.

James, C. (1999), *Global Review of Commercialized Transgenic Crops: 1999*, New York: ISAAA.

Jazairy, I., M. Alamgir and T. Panuccio (1992), *The State of World Rural Poverty: An Inquiry into its Causes and Consequences*, New York: New York University Press.

Jefferson, R.A. (1994), 'Apomixis: A social revolution for agriculture', *Biotechnology and Development Monitor*, 19, 14–16.

Jefferson, R.A.J. (1993), 'Beyond model systems: New strategies, methods and mechanisms for agricultural research', *Annals of New York Academy of Sciences*, 700, 53–73.

Joly, P.-B. and S. Lemarie (1998), 'Industry consolidation, public attitude and the future of plant biotechnology in Europe', *AgBioForum*, 1, 85–90 (see: *http://www.agbioforum.missouri.edu*).

Jorde, T. and D. Teece (1992), Antitrust, Innovation and Competitiveness, New York: Simon and Schuster.

Kalaitzandonakes, N. (1998), 'Biotechnology and the restructuring of the agricultural supply chain', *AgBioForum*, 1, 1–3.

Kindinger, P.E. (1998), 'Biotechnology and the AgChem industry', *AgBioForum*, 1, 74–5 (see: *http://www.agbioforum.missouri.edu*).

Komen, J. and G. Persley (1993), 'Agricultural biotechnology in developing countries: A cross-country review', ISNAR – IBS Research Report no. 2, ISNAR, The Hague.

Kryder, R.D., S.P. Kowalski and A.F. Krattiger (2000), 'The Intellectual and Technical Property Components of Pro-Vitamin A Rice (Golden Rice Tm): A

Preliminary', *Freedom-To-Operate Review*, ISAAA Briefs no. 20, Ithaca NY: ISAAA, p.56.

Lacy, W.B., L.R. Lacy and L. Busch (1989), 'Agricultural biotechnology research: Practices, consequences and policy recommendations', *Agriculture and Human Values*, 5, 3–14.

Lawler, C., R. van der Meer and J. Viseur (1998), 'Transferring EU-funded biotechnology research to European bioindustry', *Nature Biotechnology*, 16, 494.

Leisinger, K. (1996), 'Ethical and ecological aspects of industrial property rights in the context of genetic engineering and biotechnology', Basle: Novartis Foundation for Sustainable Development.

Leisinger, K. (1999), *The Socio-political Impact of Biotechnology in Developing Countries*, Basle: Novartis Foundation for Sustainable Development (see: *http://www.foundation.novartis.com/biotech.htm*).

Lesser, W. (1998), 'Intellectual property rights and concentration in agricultural biotechnology', *AgBioForum*, 1, 56–61 (see: *http://www.agbioforum.missouri.edu/agbioforum/*).

Levin, R.C. *et al.* (1987), 'Appropriating the returns from industrial R & D', Brookings Papers on Economic Activity, 783.

Lipton, M. (1999), 'Reviving the stalled momentum of global poverty reduction: What role for genetically modified plants?', Crawford Memorial Lecture, 28 October, CGIAR International Centers Week, Washington, DC.

Love, J. (1999), 'Health Care and IP: The Orphan Drug Act', consumer project on technology (see: *http://www.cptech.org*).

Mansfield, E. (1986), 'Patents and innovation: An empirical study', *Management Science*, 32, 173.

Marshall, A (1997), 'Millennium signs away plant kingdom to Monsanto', *Nature Biotechnology*, 15, 1334.

Mascarenhas, D. (1998), 'Negotiating the maze of biotech "tool patents"', *Nature Biotechnology*, 16, 1371–2.

Masood, E. (1999), 'Europe bids to pull US patent law into line with first-to-file system', *Nature*, 397, 457.

Massieu, Y. (1998), 'ELM: A new global player in the vegetable market', *Biotechnology and Development Monitor*, 34, 51–73.

McGowan, K. (1997), 'More adventures', *Nature Biotechnology*, 15, 824.

Meagher, L.R. and F. Bolivar (1998), 'Changing university roles in the century of biotechnology', *Nature Biotechnology*, 16, 598–9.

Mercoiret, M.R., F. Goudaby, F. Ndiame and J. Berthome (1990), 'The role of farming organizations in developing and spreading innovations', project report for CADEF, ISRA, CIEPAC, ENEA, DSA-CIRAD.

Merges, R.P. and R.R. Nelson (1990), 'On the complex economics of patent scope', *Columbia Law Review*, 839.

Miller, H.I. (1999), 'The real curse of Frankenfood', *Nature Biotechnology*, 17, 113.

Moore, J.H. (1998), 'Transaction costs, trust and property rights as determinants of organizational, industrial and technological change: A case study in the life sciences sector', in *Proceedings of Economic and Policy Implications for Structural Realignment in Food and Agriculture Markets: A Case Study Approach* (see: *http://www.ag.uiuc.edu/famc*).

Moris, J. (1991), *Extension Alternatives in Tropical Africa*, London: ODI.

Nuffield Council (2000), *Genetically Modified Crops: The Ethical and Social Issues*, London: Nuffield Foundation.

Ochave, J.M.A. (1997), 'Barking up the wrong tree: Intellectual property law and genetic resources', *BINAS News*, vol. 3 (see: *http://www.binas.unido.org/binas/*).
Pavitt, K. (1997), 'Do patents reflect the useful research output of universities?', SPRU Working Papers Series, no. 6, Brighton.
Pinstrup-Anderson, P. and J.I. Cohen (2000), 'Modern biotechnology for food and agriculture: risks and opportunities for the poor', in G.J. Persley and M.M. Lantin (eds), *Agricultural Biotechnology and the Poor*, Washington, DC: CGIAR, pp. 159–72.
Pinto, Y.M., R.A. Kok and D.C. Baulcombe (1999), 'Resistance to rice yellow mottle virus (RYMV) in cultivated African rice varieties containing RYMV transgenes', *Nature Biotechnology*, 17, 702–7.
Pray, C., S. Ribeiro, R. Mueller and P. Rao (1991), 'Private research and public benefit: The private seed industry for sorghum and pearl millet in India', *Research Policy*, 20, 315–24.
Pray, C.E. and D. Umali-Deiniger (1998), 'The private sector in agricultural research systems: Will it fill the gap?', *World Development*, 26, 1127–48.
Pray, C.E., D. Ma, J. Huang and F. Qiao (2000), 'Impact of Bt Cotton in China', paper presented at '4th International Conference on the Economics of Agricultural Biotechnology', August 24–8, Ravello, Italy.
Qaim, M. (1999a), 'Assessing the impact of banana biotechnology in Kenya', ISAAA Brief no. 10, New York, (*www.isaaa.org*).
Qaim, M. (1999b), 'The Economic Effects of Genetically Modified Orphan Commodities: Projections for Sweetpotato in Kenya', ISAAA Brief no. 13, New York (*www.isaaa.org*).
Rabobank (1994), 'The World Seed Market: Developments and strategy', Rabobank, Netherlands.
Ratner, M. (1998), 'Competition drives agriculture's genomics deals', *Nature Biotechnology*, 16, 810.
Reddy, P. (1997), 'New trends in globalization of corporate R&D and implications for innovation capability in host countries: a survey from India', *World Development*, 25, 1821–37.
Renkoski, M. (1998), 'Unlike water, money can flow upstream and new food systems will make it happen', *AgBioForum*, 1, 81–4 (see: *http://www.abioforum.missouri.edu*).
Rivera, W.M. (1996), 'Agricultural extension in transition worldwide: structural, financial and managerial strategies for improving agricultural extension', *Public Administration and Development*, 16, 151–61.
Rivera, W.M. and D.J. Gustafson (eds) (1991), *Agricultural Extension: Worldwide Institutional Evolution and Forces for Change*, Amsterdam: Elsevier.
Robertson, D. (1999), 'Pharma strategies extend drug lives', *Nature Biotechnology*, 17, 220–21.
Roling, N. (1989), 'The agricultural research–technology transfer interface: a knowledge systems perspective', Linkages Theme Paper no. 6, 20–25 November, The Hague, Netherlands.
Royal Society (2000), *Transgenic Plants and World Agriculture*, report prepared under the auspices of the Royal Society of London, US National Academy of Sciences, Brazilian Academy of Sciences, Chinese Academy of Sciences, Indian Academy of Sciences, Mexican Academy of Sciences and the Third World Academy of Sciences.
Salamini, F. (1999), 'North–South innovation transfer', *Nature Biotechnology*, Supplement 17, BV11–12.

Scarborough, V. (1996a,b,c), 'Farmer-led approaches to extension', AGREN Papers nos 59a,b,c, ODI, London.

Scotchmer, S. (1991), 'Standing on the shoulders of giants: Cumulative research and the patent law', *Journal of Economic Perspectives*, 29.

Sehgal, S. (1996), 'IPR driven restructuring of the seed industry', *Biotechnology and Development Monitor*, 29, 18–21.

Sen, A. (1981), *Poverty and Famine: An Essay on Entitlement and Deprivation*, Oxford: Clarendon Press.

Senker, J., C. Enzing, P.-B. Joly and T. Reiss (2000), 'European exploitation of biotechnology – do government policies help?' *Nature Biotechnology*, 18, 605–8.

Shimoda, S.M. (1998), 'Agricultural biotechnology: Master of the Universe?', *AgBioForum*, 1, 62–8 (see: *http://www.agbioforum.missouri.edu*).

Spillane, C. (1999), 'Recent developments in biotechnology as they relate to plant genetic resources for food and agriculture', Background Study Paper no. 9, FAO Commission on Genetic Resources, Rome.

Spillane, C. (2000), 'Can agricultural biotechnology contribute to poverty alleviation?', CABI AgBiotechNet, March (*http://www.agbiotechnet.com*).

Spillane, C. and A.M. Thro (2000), 'Farmer participatory research and pro-poor agricultural biotechnology', *Den Ny Verden*, vol. 1 (in Danish) Centre for Development Research, Copenhagen (*http://www.cdr.dk*).

Swaminathan, M.S. (1991), *Reaching the Unreached: Biotechnology in Agriculture – A Dialogue*, Madras: Macmillan India Ltd.

TAC (1998), *Report of the CGIAR Panel on Proprietary Science and Technology*, Document SDR/TAC:IAC/98/7.1, Washington, DC: CGIAR.

Teng, P.S., M. Stanton and M. Roth (2000), 'The changing private sector investment in rice', in *Proceedings of a Workshop on the Impact on Research and Development of Sui Generis Approaches to Plant Variety Protection in Developing Countries*, 1–18 February, Los Banos, Philippines: IRRI.

Thaim, M. (1998), 'Transgenic virus resistant potatoes in Mexico: Potential socio-economic implications of North–South transfer', ISAAA Brief no. 7, New York (see: *http://www.isaaa.org*).

Thayer, A. (1995), 'Scope of agricultural biotechnology patents sparks debate', *Chemical & Engineering News*, 21 August, pp. 12–13.

Thro, A.M. and C. Spillane (2000), 'Biotechnology assisted participatory plant breeding: Complement or contradiction?', working document, CIAT, Cali, CGIAR Systemwide Initiative on Participatory Research and Gender Analysis.

Tripp, R. and D. Byerlee (2000), 'Public plant breeding in an era of privatisation', ODI Natural Resource Perspectives no. 57, June, ODI, London.

UNDP (1999), *Human Development Report*, New York: UNDP.

UNICEF (1997), *The State of the World's Children 1997, Focus on Child Labour*, Oxford: Oxford University Press.

Wambugu F. (1996), 'Control of African sweet potato virus disease through biotechnology and technology transfer', in J. Komen, J.I. Cohen and O. Zenda (eds), *Turning Priorities into Feasible Programs: Proceedings of a Policy Seminar on Agricultural Biotechnology for East and Southern Africa*, The Hague: ISNAR, pp. 75–81.

White, B. (1996), 'Globalisation and the child labour problem', *Journal of International Development*, 8, 829–39.

WHO (2000), 'Assessing the global needs for vaccine research and development.

Results of a Joint GAVI/WHO meeting, Geneva, 4–5 November, 1999', WHO, Geneva (*www.vaccines.who.int/vaccines-documents/*).
Ye, X., S. Al-Babili, A. Kloti, J. Zhang, P. Lucca, P. Beyer and I. Potrykus (2000), 'Engineering the provitamin A (beta-carotene) biosynthetic pathway into (carotenoid-free) rice endosperm', *Science*, 287, 303–5.

PART II

A Case Study on Terminators: the Impacts of Biotechnologies on Benefit Distribution

5. The impact of terminator gene technologies on developing countries: a legal analysis

William W. Fisher*

To assess the likely effects of 'terminator' technologies on the societies and economies of developing countries, one needs first to understand the roles played by intellectual property rights (IPRs) in those countries. Accordingly, this preliminary memorandum is divided into two parts. The first briefly reviews and evaluates the types of IPRs currently available for new plant varieties. The second evaluates the potential roles of 'terminator' genes as supplements or substitutes for those IPRs.

It bears emphasis that this chapter deals exclusively with the potential function of terminator technologies in the protection of intellectual property. Assessment of the health and ecological effects of that technology is left to other writers.

BACKGROUND

Legal Protection for New Plant Varieties

The character and amount of intellectual property protection accorded genetically modified plants throughout the world are currently in flux. Broadly speaking, protection is most generous in the USA, is slightly less generous in other parts of the developed world and is substantially less generous in most developing countries.

In the USA, new plant varieties created through genetic engineering are subject to protection under three statutory systems: the Plant Protection Act (covering asexually reproducing new plants varieties), the Plant Variety Protection Act (covering sexually reproducing new plants varieties) and the Patent Statute. The most generous of these three regimes is the last. Until the 1980s, it was not clear that the Patent Statute covered living things – and

* I am grateful for the comments and suggestions of Calestous Juma and Professor Susan McCouch.

plants in particular. The *Chakrabarty* and *Hibberd* decisions removed the doubts on that score. Since then, the developers of new plant varieties, and specifically of new varieties created through genetic engineering, have relied whenever possible on patent law.

The development of the law in Europe loosely paralleled the development of the law in the USA. Beginning in the 1940s, several European countries adopted special statutes providing protection for new plant varieties. The formation in 1961 of the Union pour la Protection des Obtentions Végétales (UPOV) resulted in the repeal of these statutes and their replacement with a uniform system of 'Breeders' Rights'. The UPOV has been amended several times (most notably in 1991), and the contours of its successive iterations vary. The protections it provides to plant breeders roughly resemble patent law – with three significant exceptions. First, it provides no protection for novel methods of developing plants. Second, it provides less protection than patent law for 'equivalent' plant varieties. Third, the 1978 version of UPOV (although not the 1991 version) contains an exemption for 'farmers' rights'; farmers, in other words, are permitted to reuse seeds produced by protected crops.[1] The formation of the European Patent Convention (EPC) in 1973 and the subsequent reform of individual countries' patent laws held out to European plant breeders hope for even more generous legal protection. However, the extent to which plant varieties (and plants themselves) are entitled to coverage under the EPC has varied unpredictably ever since.[2]

Until quite recently, most countries in the developing world lacked any intellectual property systems applicable to genetically altered plants. Pressure exerted by TRIPS (the 1994 GATT Agreement on Trade-Related Aspects of Intellectual Property Rights) is gradually changing that situation. Article 27 of TRIPS imposes on all signatory countries an obligation to 'provide for the protection of plant varieties either by patents or by an effective sui generis system or by any combination thereof'.[3] The most important template for 'an effective sui generis system' has been UPOV (discussed above). Compliance by developing countries with their obligations under TRIPS has been slow and uneven,[4] for reasons aptly summarized by Professor Johnson Ekpere at the recent Global Conference on Biotechnology. In many developing countries, the politicians who negotiated and signed the Agreement are no longer in power. Establishing systems of the sort required by TRIPS will be costly. Finally, the benefits conferred upon developing countries of a vigorous intellectual property regime are far from obvious.[5] These circumstances make it likely that many developing countries will fail to meet their TRIPS deadlines. Whether threats of trade sanctions by the USA will be sufficient to overcome their reluctance to comply with the Agreement remains to be seen.[6] The bottom line is that we can expect that the levels of intellectual property protection for genetically modified plants will remain lower – and perhaps dramatically lower – in developing countries than in the USA and Europe for some time to come.

The Merits and Demerits of IPRs in Plants

The standard justification for intellectual property protection for new plant varieties is the same as the standard justification for intellectual property in general: knowledge of how to produce and replicate superior plant varieties is a classic 'public good'. Unlike most goods and services, it can be used and enjoyed by unlimited numbers of persons without being 'used up'. Partly as a consequence, denying access to such knowledge to persons who have not paid for the right to enjoy it is impracticable. These conditions create a risk that new plant varieties that would be worth more to consumers than the costs of creating them will not be created because the monetary incentives for doing so are inadequate. Intellectual property protection mitigates this source of economic inefficiency. By forbidding the non-permissive replication of new plant varieties, the law increases the incentives for firms in the private sector to develop and market those varieties.

It has long been recognized, however, that pursuit of this strategy has three potential drawbacks. First, by deliberately conferring monopoly power upon the creator of the new plant variety, intellectual property rights foster 'deadweight losses'. Consumers (that is, farmers) who value the seeds of the new plant variety at more than the marginal cost of producing them but less than the monopoly price charged by the patent owner will not buy them. The net result, in economic terms, is a sacrifice of the consumer surplus that would have been reaped by the excluded consumers had the variety been made available to them at its marginal cost. This effect is traditionally represented graphically as in Figure 5.1.

Rectangle 1 represents the profit enjoyed by the owner of the intellectual property as a result of being empowered to engage in monopoly pricing. (It is this profit that provides the incentive to develop the plant variety.) Triangle 2 represents the consumer surplus reaped by farmers (and consumers of their crops) who are able and willing to purchase the seeds of the new plant variety at the monopoly price. Triangle 3 represents the social loss with which we are currently concerned – the consumer surplus that would have been reaped by consumers represented by the line *FH*, which is sacrificed because those consumers have been 'priced out of the market'. The human reality that lies behind this triangle may be as mild as a forgone opportunity to taste an especially delicious variety of fruit or as severe as the starvation of impoverished persons deprived of access to a new, more productive or more disease-resistant strain of rice.

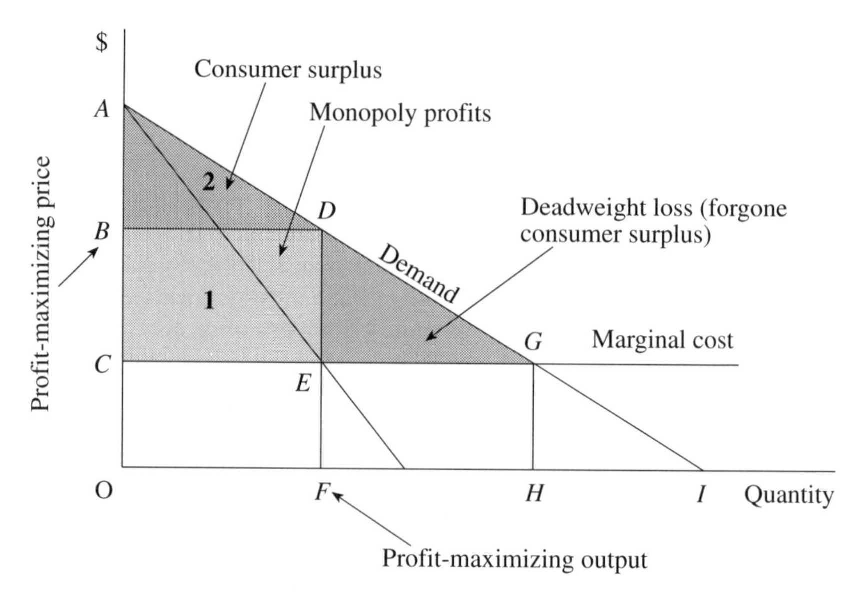

Figure 5.1 Economic effects of profit-maximising behaviour by the patentee of a new plant variety

The second of the familiar drawbacks of intellectual property protection – in general and for new plant varieties in particular – is commonly known as 'rent dissipation'. The lucrative prize of patent protection, in combination with the 'winner-take-all' regime that characterizes most intellectual property systems, tends to attract an inefficiently large number of private firms into the race to develop new varieties. The result is a waste of scientific and social resources. Scholars continue to debate the relative merits of alternative techniques of mitigating such 'rent dissipation'. No one, however, suggests that it is possible to eliminate this effect altogether.[7]

Third, intellectual property regimes tend to foster the concentration of economic power and to exacerbate disparities of wealth among the nations of the world. Both of these tendencies are evident in the context of genetically modified plants. Considerable consolidation among the firms that supply seeds worldwide has occurred since patent-protected innovations became important.[8] In addition, almost all of the relevant plant patents are now held by firms located in developed areas of the world, resulting in a flow of revenue from the poorest to the richest countries.[9]

Do the benefits of intellectual property protection justify incurring the drawbacks associated with it? In other words, do the socially valuable incentive effects of IPRs more than offset (a) the associated deadweight losses through monopoly pricing, (b) welfare losses through rent dissipation, and (c) the injuries associated with oligopolistic industrial structures and increased inequality of wealth? With respect to many fields of technology, the answer is not clear; scholars continue to debate the question whether, on balance, we would be better off without any IPRs at all.[10] With respect to biotechnology, however, there is general agreement that IPRs are essential to the industry.[11] Anecdotal evidence suggests that this overall judgment concerning the importance of IPRs to the biotechnology industry as a whole holds true with respect to the more specific issue of genetically modified crops. In the absence of intellectual property protection (for example, in most developing countries), private sector firms have been highly reluctant to invest significant resources in developing new varieties of self-pollinating plants. The only major exception to this pattern has involved crops (such as corn and maize) that lend themselves to improvement through hybridization. The fact that such hybrids are only highly productive in the first generation, combined with the ability of private firms to keep secret the highly refined stocks from which the hybrids are produced, has created a kind of 'natural' intellectual property protection sufficient to attract investors and stimulate innovation.[12] But self-pollinating crops (such as rice or wheat) that do not lend themselves to this breeding and marketing strategy have traditionally not attracted significant private investment in creating superior varieties.

In sum, if our objective is to stimulate the development of new plant varieties through genetic engineering,[13] we should support the establishment and effective enforcement of an intellectual property system, in developing countries as well as in developed countries. That general guideline, however, does not imply that we should apply automatically to genetically modified plants all aspects of the patent system currently in force in the USA. Strategic adjustments in the patent system – or the adoption of a sui generis form of plant protection (of the sort permitted by TRIPS) – may make it possible to reap the social and economic advantages of intellectual property while limiting the concomitant drawbacks.

'TERMINATOR' TECHNOLOGY

Technology protection systems enable the developers of genetically modified plant varieties to alter the characteristics of the plants while they are growing. The best known of these systems – popularly known as 'terminator' genes – render the seeds produced by genetically modified plants sterile. In other words, seeds containing the gene will produce only one generation of plants; farmers cannot, by saving the seeds generated by those plants, produce additional crops in future years.

The technology necessary to produce this effect is already sufficiently advanced to have garnered a patent in the USA. Less well developed but clearly foreseeable are so-called 'second-generation' terminator technologies. These include systems that will make it possible for farmers, by applying innocuous but specific chemicals to genetically modified crops, to 'turn on' and 'turn off' their novel characteristics (such as resistance to specific insects). As yet, most of the debate surrounding 'terminators' concerns first-generation technologies – and this chapter will similarly concentrate on the basic form. However, is it worth bearing in mind that more sophisticated systems are already on the horizon.[14]

The most controversial of the potential functions of 'terminator' genes is their capacity to serve as substitutes for IPRs. By incorporating 'terminator' genes into their seeds, the developers of new plant varieties are able to prevent farmers from making 'copies' of those seeds. To be sure, these genes do not prevent *competitors* from using genetic engineering to produce and then sell identical seeds. To shield the developers of new varieties from such competition, intellectual property law remains necessary. However, the most direct and widespread threat to the revenues of private firms developing new plant varieties – namely, the capacity of the farmers who buy the seeds to produce unlimited copies of the plants in question – is removed by the 'terminator' technology.

In this respect, adoption of 'terminator' technology to shield new plant varieties would precisely parallel recent changes in the ways in which informational products are marketed on the Internet. For centuries, the producers of informational products have relied primarily upon copyright law to shield their creations from unauthorized copying, thus enabling the producers to engage in monopolistic pricing strategies and earn enough income to cover their costs of creation. Three characteristics of the Internet, in combination, have undermined that traditional strategy. First, digital copies of informational products can be made and transmitted with remarkable ease. Second, it is difficult to trace the persons who, without permission, copy and transmit copyrighted materials on the Internet. Third, the strongly anarchic culture of the Internet increases the frequency of such unauthorized copying. In the mid-1990s, political leaders sympathetic to the plight of the producers of informational products attempted to meet this threat by tightening copyright law on both national and international levels. In the past few years, however, their reliance on the copyright system has diminished. A growing number of producers are now relying instead on either contracts or technological protections. Thus, for example, software manufacturers now commonly use 'click-on licences' to extract from their customers agreements not to engage in a variety of unauthorized uses of the products. Even more effective in protecting the economic interest of suppliers are the increasingly common forms of encryption – cryptolopes, trusted systems, serial copy management systems, and so on. Inexpensive and largely self-enforcing, these technological protections are rapidly becoming the preferred mode of shielding informational products.[15]

The marketing of genetically modified crops seems to be following a similar trajectory. The developers of new plant varieties continue to push hard for strengthened IPRs. In recent years, however, they have come to rely increasingly upon contracts to prevent unauthorized replication of their products. For example, Monsanto commonly insists that farmers agree not to reuse seed produced by its genetically modified crops.[16] In the future, 'terminator' technology may provide the seed suppliers with technological shields closely analogous to the encryption systems already being employed by software and entertainment suppliers.

Would the use of this new technology as a substitute or supplement for intellectual property protection be socially desirable? In particular, would it be good for developing countries? No simple answer to that question is apparent. Rather, the technology would have some substantial social advantages, but also some substantial drawbacks. These are sketched below.

Advantages

The most obvious benefit of the new technology is that it would enable the developers of new plant varieties to make a profit selling seeds in countries that currently lack intellectual property protections for plants. Opportunities for such profits should encourage biotechnology firms to develop plant varieties suitable for cultivation in developing countries and then to make those varieties available to farmers. If, as some observers believe, innovations in biotechnology are essential to fuel a second 'green revolution' – which, in turn, is essential to stave off the looming crisis in world food production – terminator technology might well facilitate an increase in the quality and quantity of food available to the world's poorest people.

Second, 'terminator' technology is more than a potential substitute for IPRs; from the standpoint of the private firms, it is even better than IPRs. It is plainly superior (from the firms' standpoint) to 'sui generis' plant protection statutes styled on the UPOV, for the obvious reason that it overrides the traditional entitlement of farmers to reuse seeds produced by protected plant varieties. For two, less obvious, reasons, it is also superior (again, from the firms' standpoint) to patent protection. First, although licence agreements can enable patentees to forbid farmers to reuse seeds, those licence restrictions are notoriously difficult to enforce, even in the USA, where respect for patents is reasonably strong and where enforcement mechanisms are reasonably efficient. In developing countries, the practical impediments to effective enforcement of restrictive licences are much more severe.[17] By contrast, 'terminator' technology makes it simply impossible for farmers to reuse the seeds. Second, patent law in many countries limits in various ways (for example, through the 'patent misuse' doctrine) the kinds of licence agreements that patentees may extract from their customers. By employing 'terminator' technology, the firms can escape those restrictions. These two differences suggest that the incentive effect of terminator technology will be even stronger than that of IPRs. (Whether they give rise to unacceptable adverse side-effects will be considered below.)

Third, by discouraging arbitrage, 'terminator' technology increases the ability of seed suppliers to engage in price discrimination and other forms of precise marketing. To be sure, the technology does not eliminate arbitrage altogether; farmers who bought seeds at low prices could, instead of planting them, resell them to other farmers able and willing to spend more. But plainly, it would prevent farmers from selling to their neighbours seeds produced by first-generation crops. Enhancement of the suppliers' ability to engage in price discrimination would, in turn, have two predictable effects. First, it would increase the suppliers' potential profits. Second, it would enable and encourage the suppliers to make seeds available at lower prices to poorer farmers, thus reducing the welfare losses associated with monopoly pricing. These effects are traditionally represented graphically as in Figure 5.2.

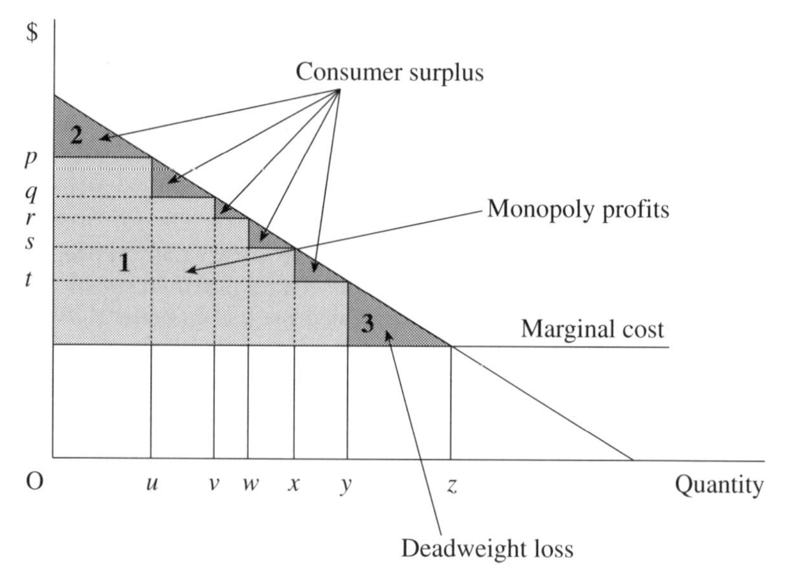

Figure 5.2 Economic impact of partial price discrimination

The essence of so-called 'partial price discrimination' is the strategy of dividing the pool of potential customers into segments and then charging the members of each subgroup what they are able and willing to spend. For example, on the simplified assumptions embodied in Figure 5.2, the seed supplier would charge the relatively wealthy and/or efficient farmers represented by line O–*U* price *p*, charge the less wealthy farmers represented by line *U*–*V* price *q*, charge farmers *V*–*W* price *r*, charge farmers *W*–*X* price *s*, and charge farmers *X*–*Y* price *t*. By adopting this strategy, the supplier would be able to increase its profits substantially. (Contrast the size of zone 1 in Figure 5.2 with the size of zone 1 in Figure 5.1, above.) At the same time, the set of farmers 'priced out of the market' would shrink. (Contrast the size of zone 3 in Figure 5.2 with the size of zone 3 in Figure 5.1, above.) Both of these effects represent improvements over monopoly pricing in the absence of price discrimination.[18] (Whether price discrimination also has unacceptable adverse side-effects we will also consider below.)

Finally, in two related respects, the use of 'terminator' technology has additional potential benefits. First, it would eliminate inefficient and inequitable cross-subsidies of long-term users by short-term users of new plant varieties. Suppose farmer A wishes to use a new variety of genetically altered rice for one year, while farmer B wishes to use it for ten years. In the absence of 'terminator' technology, it will be difficult for the supplier to differentiate among the two farmers. Consequently, they will pay the same price for the seed. Farmer B will then reuse seed from his first crop in subsequent years, while farmer A will not. The net effect is that Farmer A will pay the same amount for one year's crop as Farmer B pays for ten. Such a situation both is unfair and distorts their incentives to use new varieties. 'Terminator' technology, by contrast, would enable the supplier to lower its prices drastically, but then charge farmers each year. The beneficial result is that Farmer B would pay ten times as much as Farmer A. The second, related effect is that the use of 'terminator' technology would provide farmers with optimal incentives to experiment with new varieties. In the absence of the technology (or of effectively enforced IPRs), farmers have a strong incentive to stick with the first generation of genetically improved plants they purchase, because they could reuse the seeds produced by the first generation for nothing. If better varieties become available later, farmers will be reluctant to buy them. This distortion of their incentives would be eliminated if they had to pay for new seeds each year.

Disadvantages

The potential disadvantage of 'terminator' technology that looms largest in current debate is the danger that farmers will become dependent on

Western biotechnology companies for their supplies of seeds.[19] The companies, it is feared, will then take advantage of that situation to charge farmers exorbitant prices. The net effects will be further to impoverish farmers in the developing world – and to increase the regrettable flow of revenues from developing countries to the developed world.

A more subtle variation on this theme is the risk that the biotechnology companies will use their economic leverage to trap farmers in a cycle of economic dependency analogous to the 'crop lien' system that prevailed in the American South in the late nineteenth century.[20] In some areas, poor farmers may be unable to pay for the seeds at the start of the growing season. The seed companies will thus be tempted to provide the farmers with the seeds they need in exchange for a promise to pay the companies higher prices at the time of harvest – a promise the farmers will likely have trouble keeping. After a few repetitions of this arrangement, farmers would find themselves hopelessly in debt.[21]

Defenders of the technology protest that farmers are not as vulnerable as either of these scenarios would suggest. After all, they are not forced to use the new plant varieties. Only if the new varieties are dramatically superior to existing varieties will they purchase the new seeds. And, if the companies are foolish enough to raise the prices of the new seeds to exorbitant levels, the farmers can easily switch back to the traditional varieties.

Various circumstances suggest that these responses are not wholly persuasive and that the dangers highlighted by the critics are real. First, most arrangements between seed suppliers and farmers in developing countries will involve extreme inequalities of information and bargaining power.[22] Those conditions make it likely that the farmers will enter into deals that are not in their long-term best interest. Second, it is far from clear that seeds from the traditional non-proprietary plant varieties will remain available to the farmers indefinitely. Widespread usage of superior, proprietary, genetically engineered varieties may in one way or another 'drive out' the older types. Stocks of the older seeds may be abandoned – or may lose their fertility; farmers who produce for the market may find that customers now expect products of the new types; and so forth. If this occurred, the economic power of the seed companies would increase dramatically.

What Is to Be Done?

The substantial potential benefits of terminator technology, outlined earlier, suggest that it would be unfortunate if the seed companies were compelled – either through legal prohibitions or popular opinion – to eschew use of it altogether. On the other hand, developing countries would be wise to regulate its usage so as to avoid the dangers also sketched above. How? Set forth below are a few possibilities.

1. Manipulation of the rules of patent law is an unpromising way of seeking to prevent the technology from being used in exploitative ways. If anything, denying a patent to this technology, thus enabling many biotechnology companies (rather than just one) to use it in marketing their genetically altered seeds, would increase rather than decrease the dangers associated with it.
2. More promising would be mandatory disclosure rules, analogous to those imposed on all residential mortgage transactions in the USA. Sellers of genetically altered seeds containing 'terminator' genes would be required to explain to each customer the implications of the new technology – and, in particular, the fact that the customer would need to purchase new seeds each year. This strategy could go some distance towards reducing the information asymmetries between sellers and buyers, but would likely not eliminate altogether the risks catalogued above.
3. More promising would be price controls. A developing country could permit the sale and use within its jurisdiction of seeds containing 'terminator' genes, but only on the condition that the seller not exceed price ceilings set by an administrative tribunal. American copyright law (although not patent law) contains many such arrangements. Generally speaking, they work well in balancing the interests of innovators and consumers. A similar strategy could enable developing countries simultaneously to create incentives for the development of new plant varieties while curbing exploitative use of the resultant economic levers.
4. Finally, more elaborate regulatory mechanisms might be effective. For example, a developing country could permit the sale and use within its jurisdiction of seeds containing the terminator gene, but only on the condition that the seller agree in all future years to continue to provide the seeds to the original purchasers at the same price (adjusted to take into account inflation). Under such an arrangement, the seller could charge a higher price to new customers, but not to existing customers.[23] Such a system would shield farmers from many of the exploitative tactics predicted by the critics of the technology.

It is impossible, without knowing a good deal more about the economic and social conditions in a specific country, to determine which regulatory apparatus would work best. But the general idea should be clear enough: a sensible response to the advent of terminator technology would permit – but regulate – its use.

NOTES

1. See George K. Foster, 'Opposing Forces in a Revolution in International Patent Protection: The U.S. and India in the Uruguay Round and its Aftermath', *UCLA Journal of International and Foreign Affairs*, 3 (1998): 283.
2. Stages in the evolution of EPC protection for plants include the *Ciba-Geigy*, *Lubrizol* and *Plant Genetic Systems* decisions, and the recent Biotechnology Directive of the European Parliament and of the Council. For a review of this tortured doctrinal history (and some proposals for ending the uncertainty), see Geertrui van Overwalle, 'Patent Protection for Plants: A Comparison of American and European Approaches', *Journal of Law and Technology*, 39 (1999): 143.
3. TRIPS article 27(3)(*b*). For explications of the provision, see Klaus Bosselmann, 'Plants and Politics: The International Legal Regime Concerning Biotechnology and Biodiversity', *Colorado Journal of International Environmental Law and Policy*, 7 (1996): 111; J. Benjamin Bai, 'Comment, Protecting Plant Varieties under TRIPS and NAFTA: Should Utility Patents be Available for Plants?', *Texas International Law Journal*, 32 (1997): 139; Robert M. Sherwood, 'The TRIPS Agreement: Implications for Developing Countries', *Idea*, 37 (1997): 491.
4. Article 65 of TRIPS allows developing countries to postpone the implementation of most of the Agreement's requirements for a period of five years, and up to ten years with respect to technology fields that were previously excluded from patent protection under their domestic laws.
5. Keith Maskus (2000), *Intellectual Property Rights and Development*, London: CABI.
6. See *World Intellectual Property Report*, 13 (1999): 310.
7. For a sampling of this literature, see Partha Dasgupta, 'Patents, Priority and Imitation or, The Economics of Races and Waiting Games', *Economics Journal*, 98 (1988): 66, 74–8; Partha Dasgupta and Joseph Stiglitz, 'Uncertainty, Industrial Structure and the Speed of R&D', *Bell Journal of Economics*, 11 (1980): 12–13; Mark F. Grady and J.I. Alexander, 'Patent Law and Rent Dissipation', *Virginia Law Review*, 78 (1992): 305; Louis Kaplow, 'The Patent–Antitrust Intersection', *Harvard Law Review*; Michael L. Katz and Carl Shapiro, 'R&D Rivalry with Licensing or Imitation', *American Economic Review*, 77 (1987): 402; Edmund Kitch, 'The Nature and Function of the Patent System', *Journal of Law and Economics*, 20 (1977): 265; idem, 'Patents, Prospects, and Economic Surplus: A Reply', *Journal of Law and Economics*, 23 (1980): 205; Steven A. Lippman and Kevin F. McCardle, 'Dropout Behavior in R&D Races with Learning', *Rand Journal of Economics*, 18 (1987): 287; Glenn C. Loury, 'Market Structure and Innovation', *Quarterly Journal of Economics*, 93 (1979): 395; Robert Merges and Richard Nelson, 'On the Complex Economics of Patent Scope', *Columbia Law Review*, 90 (1990): 872; Frederic M. Scherer, 'Research and Development Resource Allocation Under Rivalry', *Quarterly Journal of Economics*, 81 (1967): 364–6; Pankaj Tandon, 'Rivalry and the Excessive Allocation of Resources to Research', *Bell Journal of Economics*, 14 (1983): 152; Brian D. Wright, 'The Resource Allocation Problem in R&D', in George S. Tolley, James H. Hodge and James F. Oehmke (eds), *The Economics of R&D Policy* (1985): 41, 50.
8. See Beth Baker, 'A new advisory panel will help USDA tackle the thorny issues raised by agricultural biotechnology', *Bioscience*, 1 June,1999.
9. See Vandana Shiva, *Biopiracy: The Plunder of Nature and Knowledge* (1997): 81.
10. For efforts to resolve the controversy, see M. Tyerman, 'The Economic Rationale for Copyright Protection for Published Books: A Reply to Professor Breyer', *UCLA Law Review*, 18 (1971): 1100; D. Abramson, 'How Much Copying Under Copyright? Contradictions, Paradoxes, Inconsistencies', *Temple Law Review*, 61 (1988): 133, 142–5; Kenneth Arrow, 'Economic Welfare and the Allocation of Resources for Invention', in *The Rate and Direction of Inventive Activity* (National Bureau of Economic Research, 1962). The large majority of scholars, however, despair of reaching any satisfactory answers to the question of the overall efficiency of the patent system. See, for example,

Arnold Plant, 'The Economic Aspect of Copyright in Books', in *Selected Economic Essays and Addresses* (1974): 58–62; Robert M. Hurt and Robert M. Schuchman, 'The Economic Rationale of Copyright', *American Economic Review*, 56 (1966): 425–6; Jessica Litman, 'The Public Domain', *Emory Law Journal*, 34 (1990): 997; Lloyd Weinreb, 'Copyright for Functional Expression', *Harvard Law Review*, 111 (1998): 1232–6; John Shepard Wiley, Jr., 'Bonito Boats: Uninformed but Mandatory Innovation Policy', *Supreme Court Review* (1989): 283; W.W. Fisher, 'Fair Use', 1739; Yen, 'Restoring the Natural Law', supra.

11. See, for example, Kevin W. McCabe, 'The January 1999 Review of Article 27 of the TRIPS Agreement: Diverging Views of Developed and Developing Countries Toward the Patentability of Biotechnology', *Journal of Intellectual Property Law*, 6 (1998), 41; Gerald J. Mossinghoff and Ralph Oman, 'The World Intellectual Property Organization: A United Nations Success Story', 160 World Affairs 104, 105 (1997).
12. See David S. Tilford, 'Saving the Blueprints: The International Legal Regime for Plant Resources', *Case Western Journal of International Law*, 30 (1998): 373, 385–7.
13. There is of course a serious question, currently being debated worldwide, whether *any* genetically modified crops are healthy and environmentally responsible. For the purposes of this preliminary report, I have put that fundamental question to one side, for two reasons: first, it lies well beyond my field of expertise; second, if the answer to the question is no, then 'terminator gene' technology is irrelevant.
14. See Charles Mann, 'Biotech Goes Wild', *Technology Review*, 102 (1999): 36.
15. These trends are described and evaluated in William Fisher, 'Property and Contract on the Internet', available at *http://www.law.harvard.edu/Academic_Affairs/coursepages/tfisher/compuls99.html.*
16. See Charles Mann, 'Biotech Goes Wild', *Technology Review*, 102 (1999): 36.
17. See John Greenwood, 'Terminator Gene Trips Alarms', *National Post*, 20 February 1999; Ethirajan Anbarasan, 'Dead-end seeds yield a harvest of revolt', *UNESCO Courier*, 1 June 1999.
18. What about the effect of price discrimination on *aggregate* consumer welfare? Notice that adoption of the pricing strategy described in the text has substantially reduced the consumer surplus enjoyed by wealthy and eager buyers (near the *Y* axis), but has made the product available to a much larger set of consumers, who are now enjoying surpluses of their own. Whether *total* consumer surplus has increased or decreased is impossible to determine. See W. Kip Viscusi, *Economics of Regulation and Antitrust*, 279–83 (1992); Michael J. Meurer, 'Price Discrimination, Personal Use and Piracy: Copyright Protection of Digital Works', *Buffalo Law Review*, 45 (1997): 845, 897–8. These various impacts are examined in considerably greater detail in Fisher, 'Property and Contract on the Internet', supra, note 15.
19. Such a situation would involve an enormous change in current practices. At present, 80 per cent of the crops in the developing world are grown each year from saved seed. Approximately 1.4 billion people depend for their primary food sources upon crops grown in this fashion. See Reungchai Tansakul and Peter Burt, 'People power vs the gene giants', *Bangkok Post*, 1 August 1999; Ethirajan Anbarasan, 'Dead-end seeds yield a harvest of revolt', *UNESCO Courier*, 1 June 1999.
20 See Ranson and Sutch, *One Kind of Freedom* (1977).
21. Plainly, this scenario would only apply to farmers who plan to sell at least some of their crops. Subsistence farmers would be immune to this particular threat.
22. For an analysis of the conceptual and practical difficulties presented by situations of this sort, see Todd D. Rakoff, 'Contracts of Adhesion: An Essay in Reconstruction', *Harvard Law Review*, 96 (1983): 1173.
23. This arrangement would closely resemble so-called 'second-generation' residential rent control statutes in the USA.

6. Impact of terminator technologies in developing countries: a framework for economic analysis

C.S. Srinivasan and Colin Thirtle

INTRODUCTION

The emerging technology for inducing sterility in seeds – or 'terminator technology' as it has popularly come to be known – has the potential to bring far-reaching changes in the seed industry and the organization of agriculture. Terminator technology alters a fundamental characteristic of seed – its self-reproducing nature – and threatens to change agricultural practices that have been in vogue for centuries. Seldom does an innovation have the potential to alter a product market in such a fundamental way. Given its potential impact, it is understandable that terminator technology has attracted a virulent and polemical response from a range of quarters. It has been seen as an unethical or immoral technology that threatens the livelihood of millions of farmers, especially resource-poor farmers in developing countries. It is viewed as a technology that needs to be stopped in its tracks, or discarded completely, if its adverse consequences are to be avoided. Such a response has discouraged a more dispassionate economic analysis of the technology. Such an analysis must attempt to identify the economic forces that have induced the development of this technology. There may be insights to be gained in viewing terminator as an induced technological response to the inadequacies and weaknesses of existing intellectual property rights (IPR) institutions. The technology has important consequences for the implementation, enforcement and duration of intellectual property rights for plant varieties. Consequently, the technology could have a significant impact on the appropriability of returns to investment in plant breeding and the level of investment in the development of new plant varieties. The potential impact on innovation in the plant-breeding sector must be taken into account when a balance sheet for the technology is drawn up. The current intense debate about the technology has obscured the fact that the choice for developing countries may not simply lie in accepting or rejecting the technology altogether. Other strategic responses, involving the regulation of the application of the technology, may be available that need to be explored.

THE TECHNOLOGY

An understanding of the complex biotechnological processes involved in terminator technology is not readily accessible for non-technical people. The details of how these processes work are contained in lengthy technical patent documents, quite difficult to comprehend even for someone skilled in the art. Ironically, while seed multinationals have been severely criticized in their attempts to promote the technology, the origin of the controversy can be traced to a patent granted jointly to the US Department of Agriculture and Delta & Pineland Company in 1996. The technology involves the following components:

1. techniques for development of seeds that grow normally when first planted by farmers, but which produce sterile seeds at the end of the cycle, that is at harvest;
2. techniques which allow companies to produce seed for sale to farmers that grow normally;
3. techniques that ensure that other product characteristics remain unaffected. This implies that terminator technology should come into play at a late stage in seed development.

From the economic point of view, the most important implication of this technology is that farmers cannot save seeds from their crops: they have to buy fresh seeds from the seed companies every year. It must be noted that this technology may be relevant only for self or open pollinated varieties; in the case of hybrids, farmers generally buy seed every year because of the loss in yield (owing to the loss of hybrid vigour) when second-generation seeds (F2) are used. There is no incentive for seed companies to put this technology into hybrid varieties, as there is already a mechanism to ensure repeat purchases of seed by farmers. However, a quick survey of the terminator literature shows that the development of the technology is not oriented only towards seed sterility. The same technology can be used to switch on or off specific traits in the seeds of a variety. Certain traits can be rendered dormant. These dormant traits may be expressed only when the seeds are used in conjunction with certain proprietary chemicals. The technology opens up new possibilities for companies to bundle together seeds and other inputs.

It is not clear in which crop species the implementation of terminator technology will ultimately prove feasible. However, if terminator emerges as a truly generic technology, capable of being implemented in a whole range of crops species and varieties, it may become the biotechnologist's vehicle of choice for the delivery of innovations. All biotechnological innovations (such as transgenic varieties resistant to insects) would be delivered to the market bundled with terminator. Even the availability of existing varieties could be affected if companies choose to sell them only in the terminator version.

It appears that, unlike genetically modified varieties that offer agronomic benefits to farmers, terminator technology offers no such benefits to them. The technology is aimed at the elimination of loss of revenue to seed companies owing to the use of farm-saved seed. It is also aimed at facilitating the joint sale of different types of inputs and price discrimination in the seed market. If the technology has emerged, not for agronomic reasons, but rather as a marketing device, it becomes necessary to explore the economic rationale for its emergence.

PLANT BREEDING AND INTELLECTUAL PROPERTY RIGHTS

It is generally recognized that the growth of agricultural production has become critically dependent on yield increases primarily based on the development of new high-yielding varieties. While, traditionally, the development of new varieties was the result of informal innovation by farmers over several generations, in the last 125 years this process has been greatly accelerated by scientific plant breeding, which has emerged as one of the most important areas of agricultural research. Its distinguishing feature is the *planned incorporation* of specific desirable traits in new varieties using the range of *techniques and information* available to plant breeders (OECD, 1993). Modern plant breeding not only takes place in a different institutional context, but this process of evolving new varieties is fundamentally different from the farmer's process that relies on careful selection from randomly occurring mutations in nature.[1]

The transformation of plant breeding into an organized scientific activity raises questions of incentives for research. The two important questions are (a) whether any form of intellectual property rights can be made applicable to plant breeding, and (b) whether the market mechanism will produce an optimal level of investment in plant breeding. In the case of industrial products the concept of intellectual property rights as a device for promoting innovation has been around for more than 500 years

(Machlup, 1958). The Paris convention for the protection of industrial property dates back to 1860. But the idea of applying IPRs to plant varieties took another 100 years to emerge. An important reason for this was that, until recently, the prevailing paradigms of IPRs precluded their application to living materials. However, the most important difficulty in applying IPRs to plant varieties arose from the difficulties in segregating[2] and appropriating benefits from their use (Swanson, 1997). This difficulty arose directly from the self-reproducing nature of seed. Once a plant breeder released a new variety to farmers through the sale of seed, he had no further control over the use of the variety, as farmers could multiply it themselves. The benefits that the breeder could appropriate by the direct use of his new variety would be only a very small fraction of the total social benefit that could be derived if his variety were to be widely diffused to farmers. The vast discrepancy between the benefits that could be appropriated privately by the breeder and the total social benefits implied that the market mechanism would fail to produce a socially desirable level of investment or effort in plant breeding. The market mechanism would invariably lead to underinvestment in the development of new varieties.

The conventional response to the problem of market failure, in developed and developing countries, has been to provide investment through public sector research, as evidenced by the establishment of land grant universities in the USA, public agricultural research institutions in Europe and the large national agricultural research systems (NARS) in several developing countries. The contribution of public research systems to the development of improved varieties has been spectacular. But financial support for public sector agricultural research has suffered a serious setback since the 1980s in both developed and developing countries (Alston *et al.*, 1998). Alston *et al.* report that in real terms, between 1945 and the mid-1970s, in most developed countries, public expenditures on agricultural R&D grew more rapidly than in the rest of the post World War II period. Then, in the mid-1970s, rates of growth in public R&D outlays slowed quite markedly and, in the 1980s, public R&D expenditures generally stagnated or declined. In the 1990s, however, public R&D expenditures recovered or began to increase again but at more modest rates of growth than in the 1960s or 1970s. This slowdown in the growth of R&D agricultural spending has been a feature of developing countries as well. Maintaining previous levels of growth in public research expenditure may prove infeasible in both developed and developing countries. The fiscal constraints of the public sector, the emerging role of biotechnology in plant breeding (that calls for a magnitude of investment only very large players can afford), changing ideas about the applicability of IPRs to living materials and scientific advances that facilitate the enforcement of IPRs for seeds have brought into sharp focus the question of IPRs for new plant varieties to provide incentives for private sector research.

If private investment in plant breeding has to be encouraged, the problem of appropriability has to be addressed. This implies what some authors (for example, Kloppenburg, 1988) have called the 'commoditization' of seed. This thrust towards commoditization has probably been the most important factor in the emergence of the concept of IPRs over plant varieties. Successful commoditization requires that institutional barriers be placed on the self-reproducing characteristics of seeds – and this is precisely what plant variety protection legislations attempted to do. Scientific progress itself facilitated commoditization. For instance, the emergence of hybrids provided a technological solution to the problem of appropriability. The emergence of hybrids encouraged the participation of the private sector in plant breeding for some crops, such as maize, in the USA. PVP (plant varietal protection) legislation can be seen as an attempt to address the problem of appropriability for self or open pollinated varieties. The operationalization of PVP was facilitated by the following developments.

1. The development of systematic botany, which permitted the unique description of varieties, based on morphological characteristics. This enabled a particular case of plant variety rights to be distinguished from any other case. Better breeding techniques made it possible to breed genetically uniform varieties. This facilitated the identifiability of individual varieties, which was a key element for the application of IPRs.
2. The increase in farm size and a decline in the number of farmers in developed countries, which drastically reduced the transaction costs of enforcing IPRs in relation to farmers.
3. The development of molecular techniques for identification of varieties and their parentage that facilitated the enforcement of IPRs in relation to imitators and competitors (see Godden, 1998).

It was in the 1960s that several European countries enacted plant variety protection legislation in the context of growing private sector participation in plant breeding activities. Special laws for plant variety protection were enacted partly because of the technical difficulties in applying the patent system (designed for industrial products) to plant varieties, which were thought not to precisely reproduce themselves and whose appearance could vary depending upon the environment in which they were grown. Prior to this, the USA had enacted a Plant Patent Act (in 1930) to provide protection to varieties of plant that reproduced themselves asexually. In the

Netherlands, the Breeders Ordinance of 1941 granted a very limited exclusive right for breeders of agriculturally important species to market the first generation of certified seed. In Germany in 1953, the Law on the Protection of Varieties and the Seeds of Cultivated Plants gave breeders the exclusive right to produce seed of their varieties for the purposes of the seed trade and to offer for sale and market such seed. In the period prior to 1961, while a number of states provided limited rights to plant breeders, the criteria for grant of rights differed from state to state and even the concept of variety was not seen in the same light in all states. There was no guarantee that the rights that the state was prepared to grant to its own nationals would be extended to the nationals of other states. Where varieties were protected in one state but not in another, several distortions could result. It was the adoption of the International Convention for the Protection of New Varieties of Plants (UPOV) in 1961 that provided, for the first time, recognition of the rights of plant breeders on an international basis.

The UPOV Convention attempted to harmonize the PVP legislation of member countries. It specified uniform criteria for the protection of new varieties as *distinctness*, *uniformity* and *stability*. These criteria reflected the need for identifiability of a variety as a prerequisite for the application of IPRs. The Convention required member states to accord the same treatment to nationals of other states as they accorded to their own nationals. It also provided for certain elements of reciprocity. Importantly, it defined the scope of the breeders' rights, which extended to *production for purposes of commercial marketing of the propagating material of the new plant variety*. The UPOV Convention of 1978 and the PVP legislation of most member countries had two important features, which distinguished the protection of plant varieties from patents. The first was farmers' privilege, which acknowledged the right of farmers to use farm-saved seed. The breeders' right extended only to the production of seed for commercial marketing and consequently the use of farm-saved seed was outside the purview of the breeders' right.

The second feature was research exemption, which provided that the use of the new variety as the initial source of variation for creating other new varieties and marketing them was free; that is, it did not require the breeder's authorization. Accordingly, the protection under the 1978 Act did not give the plant breeder any rights in the genes, the underlying genetic resource, contained in his variety. The research exemption meant that a protected variety could be used in the development of other new varieties, but it also facilitated 'cosmetic breeding'. A variety which was only marginally different from a protected variety could qualify for protection as a new variety. Such 'imitation' could deprive the original breeder of a substantial part of royalties from his protected variety.

Since the 1960s, almost all developed countries have enacted PVP legislation. Several developed countries have had the legislation for more than 20 years. Empirical studies on the impact of PVP have been scarce, but there have been a few studies that have assessed the impact of PVP in the USA and Latin American countries (Perrin *et al.*, 1983; Butler and Marion, 1985; Butler, 1996; Kalton and Richardson, 1983; Kalton *et al.*, 1989; Frey, 1996; Babcock and Foster, 1991; Alston and Venner, 1998; Jaffe and Van Wijk, 1995). While PVP does appear to have facilitated private sector participation in the breeding of non-hybrid crops, the available empirical evidence suggests that the incentive effects of PVP may be fairly weak, as farmers' and researchers' exemptions appear to constrain the appropriation of returns. PVP has not brought the appropriability of returns from investment in self-pollinated crops anywhere near the levels reached with hybrid crops. PVP does appear to play an important role in facilitating important changes in the institutional framework for agricultural research. It forces a reappraisal of the rationale for public sector intervention in plant breeding research. This can lead to a redefinition of the role of the public sector and its relationship with the private sector.

TACKLING THE PROBLEM OF APPROPRIABILITY

There have been several attempts to improve the appropriability of returns permitted by PVP laws. A comparison of the 1991 UPOV Convention with the 1978 UPOV Convention clearly brings out the nature of these efforts. The 1991 UPOV Convention extends the right of the breeder to all reproduction of the seed of the protected variety, but individual member states may provide for farmers' privilege in their laws as an exception. The farmers' privilege, therefore, becomes available only as an exception to the breeder's right. In countries where legislation has continued to provide for farmers' exemption (as an exception) it may not have made much of a difference in practice, but an important change of principle was involved. To facilitate better enforcement of the rights of the breeder, the 1991 UPOV Convention extended the rights of the breeder to the harvested material (including whole plants) in cases where the breeder had not been able to exercise his right in relation to the propagating material. Member states can optionally extend the right of the breeder even to products made from the harvested material in certain circumstances. An attempt was made to tackle the loss to breeders through 'cosmetic breeding' by introducing the concept of 'essential derivation'. The right of a breeder was extended to all varieties that could be defined as being 'essentially derived' from the protected variety. While the definition of what constitutes an essentially derived variety is controversial and is still unsettled, the concept is clearly intended to make PVP as close as possible to patents.[3] The 1991 Convention also seeks to improve appropriability for breeders by extending protection to all species and by increasing the duration of protection.

The changes in the UPOV Convention have been reflected in the changes in national laws. In almost the whole of the EU, farmers' privilege no longer exists except in the case of small farmers. Other farmers have to pay a royalty to the breeder (or the PVP title-holder) even when they use farm-saved seed of a protected variety. In the USA, the PVP Act originally provided that farmers could sell farm-saved seed to other farmers. The sale of such farm-saved seed became fairly extensive and came to be known as 'brown-bag' sales. While such sales may seem innocuous, their aggregate effects cut severely into seed companies' profits. In 1990, Pioneer Hi-Bred, one of the world's largest seed companies, decided to cease production of a variety of winter wheat in Kansas when it discovered that only 8 per cent of the variety grown in Kansas had been raised from seed actually purchased from Pioneer. The secondary brown-bag market had swallowed 92 per cent of Pioneer's market share. The right to sell farm-saved seed was circumscribed in the USA by a series of judicial decisions (Goss, 1996). In the *Asgrow* v. *Winterboer* case, the US Supreme Court held that a farmer could sell only the amount of seed that he would need to replant his own acreage. The 1994 amendments to the PVP Act removed the provision for sale from the farmers' exemption altogether. Farmers could now save seed only for replanting their own land. This measure was recognition of the impact of brown-bag sales on the seed industry.

The seed industry has also been using various other techniques to improve appropriability, such as the use of purchase contracts and label notices. A purchase contract specifically prohibits a farmer from using the harvested seed of a protected variety for replanting or for selling for breeding or variety improvement purposes. Sometimes an attempt is made to impose restrictions on the use of seed by affixing notices on the label of a product. The purchaser is deemed to agree to the restrictions when he opens the packet and uses the seed. The enforceability of such purchase contracts and label notices, especially when the product is covered by PVP and not by patents, is not very clear and will probably have to be decided by judicial decisions. But it is clear that, in the quest for better appropriability, seed companies have been exploring a range of measures. Such measures have become extremely important in the context of the sale of transgenic or genetically modified varieties. These varieties are covered by utility patents rather than by PVP. For making available the seed of transgenic varieties to farmers, seed companies like Monsanto require each farmer to enter into an agreement with the company which prohibits the farmer from using farm-saved seed. The company can carry out physical checks or what are known as 'audits' for a period of three years after the sale to ensure that the farmer has not replanted farm-saved seed. Seed companies have also been using different pricing systems. For instance, the price of transgenic seeds comprises a basic seed price and a technology fee. In certain cases the company offers to refund a portion of the technology fee, depending on the actual benefit derived by the farmer by way of lower use of pesticides and so on (Lindner, 1999). This implies that the company can effectively charge a different price to every farmer, which approaches perfect price discrimination.

The quest for stronger protection has led to plant varieties being protected through patents in a number of countries. In the USA, plant varieties became patentable as a result of a series of judicial decisions reinterpreting the existing patent laws (*Diamond* v. *Chakrabarty* and *ExParte Hibberd* are two landmark cases) (Goss, 1996). Patents provide stronger protection because they are not subject to farmers' and researchers' exemptions. The scope of utility patents is wide because it allows a breeder to exclude others from *making*, *using* or *selling* the seeds of the patented variety. The 'doctrine of equivalents' protects a breeder from imitation of his variety. Further, a breeder is also protected against inventors who independently come up with the same invention. Patent protection is expensive to obtain. A breeder must prove several elements: that the variety is novel and useful, that it is 'enabled' (currently this requirement can be satisfied by a deposit of the variety in the USA), and that it is not an obvious improvement upon an earlier protected variety. On account of the difficulty of proving that these requirements have been met, patents are frequently more difficult to obtain and take longer to obtain than PVP certificates. In the USA there has recently been a decline in the number of PVP certificates, while utility patents for plants have maintained their rising trend.

We have seen above that PVP does not significantly improve the appropriability of returns from investment in self-pollinating varieties, even with changes in laws that provide for more stringent enforcement of IPRs. Measures to restrict the use of farm-saved seed through the use of contractual arrangements, such as those described above, may be feasible only when varieties are protected by patents rather than by PVP. Utility patents are inherently more difficult to obtain because of the criteria of 'non-obviousness' and 'usefulness' that they involve (which are not applicable to PVP). Besides, contractual arrangements can be difficult to monitor to detect infringements and costly to enforce. Under contractual arrangements, seed companies are restricted to contract remedies when a breach occurs and these remedies may cover only a fraction of the losses suffered by a vendor seed company as a result of infringement. The costs of IPR-related litigation to seed companies run into millions of dollars each year. It is against this background that terminator technology can be seen as a technological solution to the problem of appropriability. In the context of changes in PVP laws that circumscribe the use of farm-saved seed and the contractual arrangements which have come into vogue for transgenic varieties, the terminator technology represents only a better technique of enforcement of IPRs, *rather than a change of principle governing what is legitimately appropriable*. The technology has emerged because the existing institutional mechanisms for enforcement were inadequate, expensive to use and imperfect. It promises to reduce drastically the transaction costs for enforcement of IPRs vis-à-vis farmers. It is a technological response to an institutional problem.

ECONOMIC IMPLICATIONS

The impact of terminator technology on appropriation of returns from investment in a self-pollinated variety is shown in the appendix, using a simple model adapted from Scherer (1986). In the simplest variant of the model, the development of a new variety by an innovator opens up a completely new market for seeds. This happens only when farmers using farm-saved seeds of traditional varieties switch to bought-in seeds of the new variety. In the second variant, the development of a new variety opens up a new market for seeds, but the innovator's market share is affected by imitators developing very similar varieties with a certain time lag. In the third and most likely scenario, the introduction of a new variety alters the market shares between innovators and competitors and does not lead to an overall expansion of market. In all the three cases the innovator/competitor does not realize the whole of the annual quasi-rent potential from the new variety instantaneously because it takes time for a variety to diffuse to farmers and build its market share. For all the three cases, the appropriation of quasi-rents from the introduction of a new variety is assessed on the basis of different assumptions about farmer seed replacement behaviour: (a) farmers do not replace seed after adopting a new variety; (b) farmers replace seed of a new variety every year; (c) a proportion of the adopters of a new variety replace seed every year; (d) adopters of a new variety replace seed at periodic intervals (say every five years). Farmers' seed replacement behaviour makes a dramatic difference to the appropriation of returns from a new variety. Terminator technology is equivalent to the case in which all farmers who adopt a new variety replace seed every year. For certain plausible values of the parameters, the discounted value of quasi-rents to the innovator are four times higher with terminator technology compared to the case in which farmers do not replace seed at all after initial purchase of a variety. Assuming that the maximum annual potential quasi-rent from the introduction of a new variety is 100 (that is, when there are no imitators or competitors), the appropriation of returns from a new variety by an innovator under different scenarios is summarised in Table 6.1.

Table 6.1 Summary of innovator's discounted quasi-rents for different scenarios

	No imitation		With imitation	
Full replacement by farmers	396	(100)	240	(60)
No replacement by farmers	39	(9)	33	(8)
Replacement by 20% farmers	118	(30)	75	(19)
Replacement every five years by farmers	58	(15)	41	(10)

Note: Figures in brackets give the percentage to the (maximum) quasi-rents in the situation with no imitators and full replacement by farmers.

The important question is whether this dramatic difference in appropriation of returns from a new variety when farmers replace seed every year gets reflected in the resources allocated to research. The evidence from the USA suggests that there is in fact a strong link between appropriability and private plant breeding efforts, as may be seen from Table 6.2. It may be seen from the table that private plant breeding expenditures for corn, a predominantly hybrid crop, are almost four times the expenditures on a self-pollinated crop like wheat. Share of purchased seed for corn is 95 per cent while for wheat it is only 40 per cent. The growth in yields of hybrids is also significantly higher than that of non-hybrids. In another study, Thirtle (1985) estimated the contribution of biological inputs to the growth in farmers' yields in the USA after accounting for changes in fertilizer, labour, machinery and land use and allowing for substitution among inputs. It was estimated that biological inputs increased corn yields by 1.7 per cent per year, wheat yields by 1.5 per cent, soybean yields by 1.1 per cent and cotton yields by 0.5 per cent. Contribution of biological inputs to yield gain appears to increase with the appropriability of returns. Fuglie *et al.* (1996) estimate that, for crops grown with hybrid seed, seed companies appeared to capture 35–48 per cent of the value of improved seed, with the remainder going to farmers. For non-hybrid crops (wheat, soybeans and cotton) seed companies obtained even lower shares of yield gains, from 12 to 24 per cent, as shown in Figure 6.1 below. This figure shows the share of genetic yield growth for these crops captured by seed companies in the form of higher seed prices and the share kept by farmers, on the assumption that half the growth in farmer yields can be attributed to genetic improvements and the other half to other factors. For the hybrid seed crops, seed companies invested over 10 per cent of seed sales in research. For the non-hybrid seed crops, only 4–5 per cent of seed sales was reinvested in research. 'The inability to capture a larger share of the gains from breeding non-hybrid crops served as a disincentive for seed companies to invest more in research' (ibid., 41).

Table 6.2 Seed sales, private plant breeding and trends in seed prices and yields of major field crops, USA (1975–92)

Crop	Seed sales	Private plant breeding	Seed cost	Share of seed purchased	Growth in seed price	Annual growth in crop yields
	Million 1989 dollars		Dollars/acre	Per cent		
Hybrid seed						
Corn	1031	112.9	21.09	95	4.75	1.33
Sorghum	90	12.6	5.13	95	5.08	1.54
Non-hybrid seed						
Wheat	256	13.5	8.92	40	0.97	1.13
Soybean	610	24.9	12.03	73	1.92	1.23
Cotton	256	4.6	14.93	74	4.46	2.23

Source: Fuglie *et al.* (1996).

If terminator technology were to be applied to self-pollinated crops, appropriability of returns would increase dramatically and the level of research expenditures could potentially go up to the level of hybrid crops: an increase to four times the current level. The resultant increase in the rate of yield gain and the cumulative impact on production could be very substantial.

In combination with certain types of contractual arrangements discussed above, terminator technology could facilitate perfect price discrimination by seed companies. Each farmer would pay a price for seed that would be related to the benefits he derives from its use. Perfect price discrimination would allow a seed company to cream off the entire consumer surplus. But it is also a well known result in economic theory that, under perfect price discrimination, the monopolist would produce the competitive level of output and the allocation of resources would be 'efficient'. Price discrimination could result in an expansion of output.

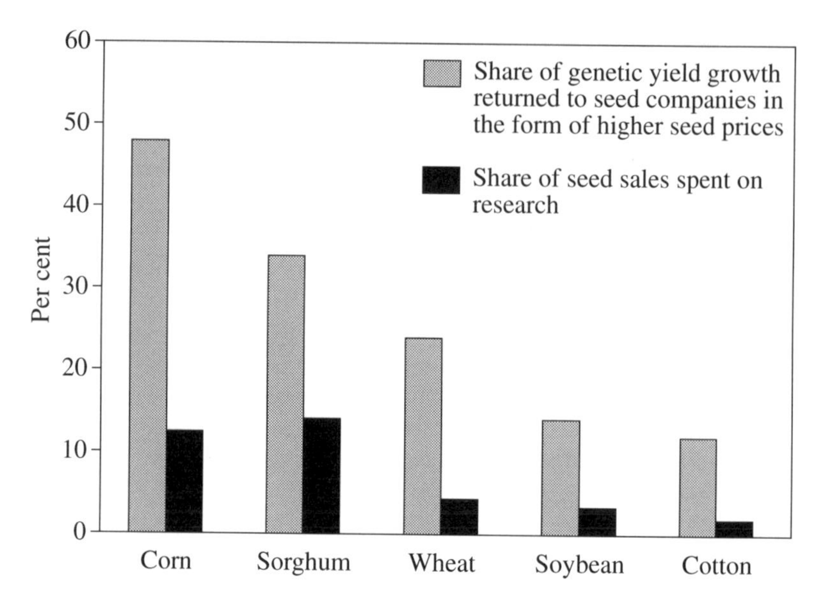

Source: Fuglie *et al.* (1996).

Figure 6.1 Appropriability and private research investment in plant breeding

Terminator technology could also open up some interesting possibilities for IPR policy. The duration of protection under PVP currently ranges from 18 to 25 years, which is longer than the duration for patents. With improved appropriability, it may be possible to reduce the period of protection to, say, ten years. This would reduce the 'deadweight' loss associated with IPRs and enable faster diffusion of varieties. However, this would be possible only if companies used PVP or patents to protect their varieties. Terminator renders PVP irrelevant for protection against unauthorized reproduction by farmers. Companies still need protection against imitation by rivals. But companies could choose to protect terminator processes through trade secrets, in which case protection would be theoretically available in perpetuity – at least until competing varieties are developed. If this happens, two key elements of IPR policy – the duration of protection and the requirement of 'disclosure' – would no longer be available to governments to influence the development and diffusion of the technology. It must be noted that trade secrets do not protect against independent discovery of a process or product, while patents do.

Theoretically, it can be argued that farmers will buy terminator seeds only if they provide higher returns even after taking into account the higher costs of seeds and other inputs that they would be compelled to buy. This assumes that farmers would have a choice and that seed markets would at least be 'contestable', if not competitive. However, the seed industry has seen consolidation through takeovers and mergers at a frightening pace. In the first wave of acquisitions and mergers in the 1960s and 1970s, many large chemical, oil and food corporations acquired many medium and small-sized seed companies. Another round of mergers in the 1980s saw many of these food, oil and chemical companies sell their interest in the seed business to agricultural chemical firms. The 1990s have seen the takeover of many small biotechnology firms by the large seed firms, and megamergers involving the seed giants (Butler, 1996). In developed countries, the new life science companies that combine crop biotechnology with agrochemical and seed production dominate the seed industry. The big six multinationals (AgrEvo, Dow, Dupont, Monsanto, Novartis and AstraZeneca), have a very substantial market share for certain crops. For instance, Monsanto could have controlled 85 per cent of the cotton seed market if its merger with Delta and Pine had gone through. If these companies all introduce terminator technology, there is a real danger that farmers may not be able to have a choice. They may not be able to choose improved agronomic characteristics without also choosing terminator technology. Local adaptation of varieties and community-based seed supply systems would cease to exist, with adverse consequences for biodiversity on farmers' fields.

The maintenance of some kind of competitive pressure, so necessary for innovation, would depend on the 'freedom to operate' that other players (such as public research systems, universities and competing companies) have. 'Freedom to operate' will depend greatly on how IPRs are applied to biotechnological processes. Terminator technology by itself confers no agronomic benefits. To be commercially successful it must be bundled with other useful agronomic traits and these traits must not be available otherwise. Increasingly, these useful agronomic traits will be ones that are engineered through biotechnology with the associated processes being patented. It is only then that companies will be able to retain exclusive use of the desirable agronomic traits. The development of new varieties in the future may require access to a number of biotechnological processes, all patented by different companies. Unless an efficient system is developed for licensing the component biotechnological processes, there could be gridlock in the development of new varieties. The scope of protection accorded to biotechnological processes needs to be carefully modulated if such a gridlock is to be avoided. There are indications that many international agricultural research centres (IARCs) are finding that they cannot work on biotechnology-based varietal improvement without licences for component processes from seed companies (Cohen *et al.*, 1998). The ability of universities and public sector institutions to licence processes often depends on their having a portfolio of patents for cross-licensing. Without the ability to cross-licence, their 'freedom to operate' would be severely restricted. The important issue is whether the public sector and other players will be able to provide improved varieties without terminator genes and offer a measure of competition.

CHOICES BEFORE DEVELOPING COUNTRIES

In acceding to the TRIPs Agreement,[4] all developing countries have committed themselves to providing some form of 'effective' IPRs for plant varieties. Several developing countries are in the process of drafting their legislation. Some are attempting to formulate a sui generis system that balances breeders' rights with farmers' rights and recognition of their contribution to the maintenance and enhancement of biodiversity. The debate about the terminator, therefore, is not about the fundamental desirability of IPRs for plant varieties but about a particular technique of enforcement. The fact that terminator technology has emerged in response to inadequacies of IPR institutions must make developing countries ponder about the kind of IPRs they need to put in place, if they really intend to encourage private investment. Stronger IPR systems may even obviate the need for development of such technologies.

An important factor, which will determine the response of developing countries to the new technology, will be the approach followed by the CGIAR[5] institutions. This will be especially true of those countries that rely on the international agricultural research centres for new varieties. The CGIAR system has apparently taken a decision that none of its institutions will use the terminator technology for the development of new varieties for developing countries in view of the potential deleterious effects of the technology. This could have important implications for the commercialization of the new technology in developing countries.

Quite independently of the CGIAR system, developing countries will have to decide on their response to the emerging technology. Very little is known about the ecological impacts of terminator varieties. The nature of interaction of terminator varieties with other crops and species is not known. It remains to be assessed whether terminator varieties are, accidentally or otherwise, capable of rendering the seeds of (adjacent) non-terminator varieties also sterile. This is a critical issue for developing countries where millions of subsistence or resource-poor farmers depend on farm-saved seed. If terminator varieties can render the farm-saved seed of non-terminator varieties sterile even in a limited number of situations, there will be a potential for a great deal of social unrest. Similarly, if terminator varieties have artificially lowered resistance to certain diseases or infestations, the unintended build-up of pathogens in the fields of both users and non-users would be a crucial issue. The large-scale use of antibiotics that currently appears to be a feature of the technology is also an important matter of concern. What developing countries require at this stage is capacity building for making detailed ecological and economic impact assessments. If improved varieties of the private sector in developed countries are going to be invariably bundled with the terminator in the future, developing countries will have to assess the trade-off between increased productivity through the use of these varieties and the retention of farmers' privileges. The existence and severity of such a trade-off will depend on the sustained capability of NARS and IARCs to produce improved varieties without the terminator technology. Their 'freedom to operate' in the brave new world of IPRs will be a critical issue. There may be a case for selective adoption of the new technology for certain crop species. If the capacity for ecological and economic impact assessment is not built up, policy will inevitably be determined by emotive, populist rhetoric.

The introduction of terminator technology in developing countries would involve interface with several elements of the regulatory framework:

1. quality control legislation,
2. varietal release and notification procedures,
3. intellectual property rights legislation,
4. policies on export and import of seed,
5. quarantine regulations,
6. approval procedures and marketing regulations for agrochemicals,
7. policy governing foreign direct investment in the seed industry.

The existing regulatory framework in developing countries may simply not be adequate to deal with terminator technology. Several elements of the regulatory framework may have to be modified substantially in a coordinated and coherent fashion to respond to the challenges posed by terminator technology. Wherever the regulatory framework is weak, developing countries will have to contend with the possibility of surreptitious introduction (or introduction through informal channels) of terminator varieties. This would be particularly important in countries where varietal registration is not mandatory for marketing of seeds, and would be more difficult to handle than introduction through the organized sector.

The policy in most developing countries has been to push for increasing the seed replacement rate (SRR) and quicker adoption of new varieties. However, even in many 'green revolution' areas such as the Indian Punjab, SRRs have been very slow to increase. This happens either because farmers cannot afford to replace seed every year or because farmers do not perceive the advantages of replacing seed or because farm-saved seed is almost equal in quality to bought-in seed. Terminator technology by its very nature can swiftly push up SRRs and also increase varietal turnover, as farmers will be more likely to change to new varieties if they replace seed every year. Developing countries have a very difficult decision to make in deciding whether terminator technology should be used as an instrument to improve SRRs and encourage varietal replacement.

Developing countries will also have to consider the impact of terminator technologies on foreign direct investment in the seed industry. There may be a tendency for the investment to flow to those countries that accept the use of this technology. Even if better appropriability generates more research investment, the location of that investment would still be an important issue for developing countries. Approval of terminator technology can greatly increase volumes for the domestic seed industry, provided it has access to the technology. The important question will be whether the domestic industry will be able to compete with terminator varieties introduced from outside. The competitive position of the domestic seed industry may depend on whether and on what terms the technology is made available to them.

The public sector in developing countries could see a major revival of its role in developing countries as the developer and producer of non-terminator varieties that may be very important for resource-poor farmers. The important question for the public sector will be whether lack of access to private sector material owing to IPRs will hamper it. Or will access to germplasm in international collections such as those of the CGIAR system be adequate for the development of competing varieties?

In assessing terminator technology, developing countries will have to address the 'green revolution' questions as well. Which farmers and which regions will be the early adopters of the new technology in developing countries? Will the adoption of the new technology be different between rich and poor farmers? Will the adoption of terminator technology follow the green revolution pattern of adoption of high-yielding varieties? What will be the implications for credit requirements? Is the technology a coercive one in the sense of compelling unwilling farmers to adopt it?

CONCLUDING OBSERVATIONS

Terminator technology has emerged as a response to the inadequacies and imperfections of existing IPR institutions. By providing a technological response to the problem of inadequate appropriability, it can be a corrective to market failure. At a time when public research systems everywhere are faced with declining budgets, and yield gains are becoming increasingly dependent on the application of expensive biotechnology, it has the potential to increase private research spending on self-pollinated crops to the level of hybrids. The resultant productivity gains may well be substantial. While better appropriability can provide additional resources for investment, there are legitimate concerns about the use of the technology. The environmental consequences of the technology can be far-reaching and complex and need to be carefully assessed. The use of the technology in an industry that is oligopolistic and highly concentrated may imply that farmers lose their right to choose and competitive pressures that are so necessary for innovation may cease. Besides, the gridlock-creating potential of IPRs can stifle, rather than promote, the development of new varieties. The scope of protection for biotechnological processes in the IPR system needs to be carefully modulated if such adverse impacts are to be avoided. But the most important lesson to be drawn from the emergence of this technology is this: in a process of market-led development, technological change will respond not only to relative factor scarcities but also to institutional bottlenecks. The development of a technology with no agronomic benefit to farmers may appear to be a huge waste of resources, but the market *will* devote resources to the reduction of transaction or appropriation costs.

Developing countries face many difficult choices in responding to this technology. If a large part of varietal improvement in the future is going to be bundled together with terminator technology, it may not be possible for them to ignore the technology altogether. Productivity is of even greater concern to them than it is to developed countries. Potential productivity gains will have to be weighed against the social costs arising from the coercive element of the technology. What they need at this stage is the capacity to evaluate its ecological, economic and social impacts, selectively apply it where appropriate and upgrade their regulatory system to deal with the technology. They will need to seriously think about appropriability and intellectual property rights.

NOTES

1. As farmers have adopted new varieties developed through agricultural research for economic reasons, the genetic uniformity of agricultural crops has increased considerably. This in turn has increased the vulnerability of agricultural crops to diseases and pests, which are constantly adapting and evolving. The development of new varieties has, therefore, become necessary not only to secure yield increases but also to keep ahead in the 'varietal relay race' against evolving pests (Plucknett *et al.*, 1987).
2. The difficulties in segregating the benefits flowing from a new plant variety arise because

productivity depends on other inputs (fertilizer, agronomic practices and so on) as well, and it is difficult to separate out the contribution of a new variety to productivity.
3. In the case of patents, the exploitation of an invention which builds upon an earlier invention requires the consent of the patent-holder of the earlier invention (during the life of the earlier invention).
4. Agreement on Trade-related Aspects of Intellectual Property Rights, which is part of the agreement establishing the WTO.
5. Consultative Group on International Agricultural Research.

REFERENCES

Alston, J.M. and R.J. Venner (1998), 'The Effects of U.S. Plant Variety Protection Act on Wheat Genetic Improvement', paper presented at the symposium on 'Intellectual Property Rights and Agricultural Research Impact' sponsored by NC 208 and CIMMYT Economics Program, 5–7 March, El Batan, Mexico.

Alston, J.M., P.G. Pardey and Vincent H. Smith (1998), 'Financing Agricultural R&D in Rich Countries: What's Happening and Why?', *The Australian Journal of Agricultural and Resource Economics*, 42(1), 51–82.

Babcock, B.A. and William E. Foster (1991), 'Measuring the Potential Contribution of Plant Breeding to Crop Yields: Flue-Cured Tobacco, 1954–87', *American Journal of Agricultural Economics*, 73 (November), 850–59.

Butler, L.J. (1996), 'Plant Breeders' Rights in the U.S.: Update of a 1983 Study', in J. Van Wijk and Walter Jaffe (eds), *Proceedings of a seminar on 'The Impact of Plant Breeders' Rights in Developing Countries' held at Santa Fe Bogota, Colombia, March 7-8, 1995*, Amsterdam: University of Amsterdam, pp. 17–33.

Butler, L.J. and B.W. Marion (1985), *The Impact of Patent Protection on the U.S. Seed Industry and Public Plant Breeding*, Food Systems Research Group Monograph 16, University of Wisconsin Madison, Madison, WI.

Cohen, J.I, C. Falconi, J. Komen and M. Blakeney (1998), 'Proprietary Biotechnology Inputs and International Agricultural Research', ISNAR Briefing Paper 39, International Service for National Agricultural Research, The Hague.

Frey, K.J. (1996), 'National Plant Breeding Study – I: Human and Financial Resource Devoted to Plant Breeding Research and Development in the USA in 1994', Special Report 98, Iowa Agriculture and Home Economics Experiment Station, Iowa State University, Iowa.

Fuglie, K., Nicole Ballenger, Kelly Day (1996), 'Agricultural Research and Development: Public and Private Investments Under Alternative Markets and Institutions', AER-735, Economic Research Service, US Department of Agriculture.

Godden, David (1998), 'Growing Plants, Evolving Rights: Plant Variety Rights in Australia', *Australian Agribusiness Review*, 6(2).

Goss, P.J. (1996), 'Guiding the Hand that Feeds: Toward Socially Optimal Appropriability in Agricultural Biotechnology Innovation', *California Law Review*, 84, 1395–1435.

Jaffe, W. and J. Van Wijk (1995), 'The Impact of Plant Breeders' Rights in Developing Countries: Debate and Experience in Argentina, Chile, Colombia, Mexico and Uruguay', Inter-American Institute for Co-operation in Agriculture and University of Amsterdam, Amsterdam.

Kalton, R.R. and P.A. Richardson (1983), 'Private Sector Plant Breeding Programmes: A Major Thrust in U.S. Agriculture', *Diversity*, 1(3), 16–18.

Kalton, R.R., P.A. Richardson and N.M. Frey (1989) 'Inputs in Private Sector Plant Breeding and Biotechnology Research Programs in the USA', *Diversity*, 5(4), 22–5.

Kloppenburg Jr., J.R. (1988), *First the Seed. The Political Economy of Plant Biotechnology 1492–2000*, Cambridge: Cambridge University Press.

Lindner, B. (1999), 'Prospects for Public Plant Breeding in a Small Country', paper presented at a seminar on 'The Shape of the Coming Agricultural Transformation: Strategic Investment and Policy Approaches from an Economic Perspective', 17–18 June, Rome.

Machlup, F. (1958), 'An Economic Review of the Patent System', Study no. 15, Sub-Committee on Patents, Trademarks and Copyrights, Committee on the Judiciary, US Senate, Washington, DC: US Government Printing Office.

Nordhaus, W. (1969), *Invention, Growth and Welfare: A Theoretical Treatment of Technological Change*, Cambridge, MA: MIT Press.

OECD (1993), *Traditional Crop Breeding Practices: An Historical Review to Serve as a Basis for Assessing the Role of Biotechnology*. Geneva: OECD.

Perrin, R.K., K.A. Kunnings and L.A. Ihnen (1983), 'Some Effects of the U.S. Plant Variety Protection Act of 1970', Economics Research Report no. 46, Department of Economics and Business, North Carolina State University.

Plucknett, Donald. L. *et al.* (1987), *Gene Banks and the World's Food*, Princeton, NJ: Princeton University Press.

Scherer, F.M. (1986), *Innovation and Growth: Schumpeterian Perspectives*, Cambridge, MA: MIT Press.

Swanson, T. (1997), *Global Action for Biodiversity: An international Framework for Implementing the Convention on Biodiversity*, London: Earthscan Publications.

Thirtle, C.G. (1985), 'Technological Change and Productivity Slowdown in Field Crops: United States, 1939–1978', *Southern Journal of Agricultural Economics*, 17, 33–42.

APPENDIX: ANALYSIS OF THE ECONOMIC RATIONALE OF TERMINATOR TECHNOLOGY

It has been argued in the text that terminator technology has emerged in response to the inadequacies of intellectual property rights (IPR) institutions for plant varieties. In particular, IPR institutions, as they operate in practice, permit only very low levels of appropriability of returns from investment in plant breeding research. The incentive effects of existing IPR institutions (such as plant variety rights) in generating research investment tend to be rather weak.

The problem of low appropriability of returns from research investment can be analysed using the framework developed by Nordhaus (1969) and Scherer (1986) for the economic analysis of patents and the allocation of R&D resources under rivalry. The framework can be adapted to plant breeding with minor modifications. The key feature of the analysis of plant breeding arises from the self-reproducing nature of seed. Most plant variety protection laws provide for 'farmers' exemption' which acknowledges the right of farmer to use farm-saved seed. This implies that, once a variety has been developed and released in the market, farmers can use seed produced from the harvest to plant their subsequent crops. They need not return to the seed company every year for fresh seed. They need to do so only after five to seven years when farm-saved seed start losing their vigour (giving lower yields) or suffer from 'genetic drift'. This has a dramatic impact on the appropriability of returns from a new variety. This argument does not apply to Fl hybrids (as farmers generally have to buy seed every year) and, therefore, the framework can also be used to illustrate the difference in appropriability between hybrids and open or self-pollinated varieties.

The other feature of plant variety rights which affects the appropriability of returns is the 'research exemption'. This provides that any protected variety can be used as a basis for developing other varieties without infringing the right of the protection holder. This affects appropriability by drastically reducing the time required by imitators and competitors to develop similar competing varieties. The speed of imitation is an important factor affecting the appropriation of benefits from a new variety.

Framework for the Seed Market

In assessing the appropriability of returns for a seed company (innovator) developing a new variety, we can conceive of three types of situations.

1. *New market*: this situation arises when a new variety opens up a completely new market for seeds. This is unlikely to arise in practice, as for any crop there will already be varieties existing in the market. Only when farmers who are completely dependent on farm-saved seeds of traditional varieties switch to bought-in seeds of modern varieties is a new market situation likely to arise.
2. *New market with imitation*: in this situation, the development of a new variety opens up a completely new market for seeds, but the innovator's market share is affected by imitators developing very similar varieties with a certain time lag.
3. *Market sharing*: this is the most likely scenario in seed markets as the area under a crop, seed use rates and seed replacement rates change rather slowly over time. Therefore the introduction of a new variety is more likely to alter market shares between the innovator and competitors rather than lead to an overall expansion of the market. During its period of technological leadership, an innovator

captures more and more of its rivals' share (and hence a larger and larger share of the quasi-rent potential of its variety). But imitators can recover market share once they have imitated.

Each of the above situations can be combined with different assumptions about farmer seed replacement behaviour: for example, 100 per cent replacement of seed every year, replacement every six years, replacement by 20 per cent of adopting farmers each year, no replacement, and so on, to ascertain the returns appropriated by the innovator from a new variety. Though the new market situation is not a likely one for the seed market (even in developing countries), it can be used to illustrate the impact of farmer seed replacement behaviour and imitation lags on the innovator's discounted stream of returns. The results from the new market situation will apply with greater force in the market-sharing situation.

Variables

Following Scherer, we define the following variables for different market situations. Subscript L denotes the innovator and subscript F denotes the imitation competitors.

General

V = potential quasi-rents (total revenues less costs) attainable from the sale of a new variety. It is initially assumed that costs and prices are constant and hence V does not depend upon the number of firms. V is the annual potential quasi-rent when *all* farmers who will eventually use the variety buy fresh seeds from the innovator in each year;

TL = innovator's new variety introduction date;

TF = imitator's product innovation date;

ρ = discount rate.

New market case

γ = penetration coefficient: the innovator will not realize the whole of the annual potential quasi-rent immediately because it takes time for a variety to diffuse to farmers and to build a market. All farmers who are eventually going to use a variety will not adopt it in the first year after introduction; γ is a coefficient that indicates the proportion of unexploited market potential captured each year. The value of γ depends upon the yield or agronomic superiority of the variety, marketing efforts and so on. The innovator moves towards realization of full market potential at the rate of 100γ per cent per year following introduction;

S_F = imitator's target market share. It is assumed that this depends on the share of the imitators in related markets and other strengths;

μ = rate at which the imitator moves toward its market share. It is assumed that the imitator moves towards full realization of its share at the rate of 100μ per cent per year following imitation;

ϵ = the imitator's permanent share erosion rate. This is assumed to vary with the lag between the introduction of a new variety by the innovator and introduction of a competing variety by the imitator. The imitator's permanent market share gets eroded at the rate of 100ϵ per cent per year of lag;

α = percentage of farmers who buy replacement seed in the current year having bought fresh seeds the previous year.

Market sharing case

δ = the innovator's (temporary) market takeover rate in market-sharing rivalry. This is assumed to be 100δ per cent per year following new variety introduction per year of lag between innovator and imitator;

β = the imitator's share recovery rate in a market-sharing rivalry (after it has introduced the competing variety).

New Market without Imitation

Full replacement by farmers

The innovator's quasi-rent in the absence of imitation and assuming farmers replace seed every year ($\alpha = 1$) is:

$$V_L = \int_{TL}^{\infty} (1 - e^{-\gamma(t-TL)} Ve^{-\alpha} dt.$$

This can be shown to be equal to

$$V_L = V\left[\frac{1}{\rho} e^{-\rho TL} - \frac{1}{\gamma + \rho} e^{-\rho TL}\right].$$

This is also the innovator's quasi-rent in the absence of imitation when the new variety is a hybrid.

No replacement by farmers

If farmers do not replace seed at all (having bought the variety once) then the market for the innovator will comprise only the new adopters every year. In this case the innovator's quasi-rent becomes:

$$V_L = \int_{TL}^{\infty} [(1 - e^{-\gamma(t-TL+1)}) - (1 - e^{-\gamma(t-TL)})] Ve^{-\alpha} - dt$$

$$= \int_{TL}^{\infty} (e^{-\gamma(t-TL)} - e^{-\gamma(t-TL+1)}) Ve^{-\alpha} - dt.$$

On simplification this yields:

$$V_L = V\frac{1}{\gamma + \rho}[e^{-\rho TL} - e^{-\gamma - \rho TL}].$$

Replacement by proportion of farmers

If we assume that a proportion α of the farmers who have adopted the seed replace their seed every year then the total market for seed for the innovator in each year becomes demand from new adopters plus replacement demand from previous adopters. The quasi-rent for the innovator can then be expressed as:

$$V_L = \int_{TL}^{\infty} [(1 - e^{-\gamma(t-TL+1)}) - (1 - e^{-\gamma(t-TL)})] Ve^{-\alpha} - dt$$

$$+ \int_{TL+6}^{\infty} (1 - e^{-\gamma(t-TL)})\alpha Ve^{-\alpha -} dt.$$

On simplification this yields:

$$V_L = V\frac{1}{\gamma + \rho}[e^{-\rho TL} - e^{-\gamma - \rho TL}] + \alpha Ve^{-\rho TL}\left[\frac{1}{\rho} - \frac{1}{\gamma + \rho}\right].$$

Replacement of seed every *n* years

If farmers replace seed, say every five years (that is, farmers who adopt the variety in the first year replace it in the sixth year) then again the total market for the innovator in each year becomes demand from new adopters plus replacement demand from previous adopters. In this case the quasi-rent for the innovator can be expressed as:

$$V_L = \int_{TL}^{\infty} [(1 - e^{-\gamma(t-TL+1)}) - (1 - e^{-\gamma(t-TL)})]Ve^{-\alpha -} dt$$

$$+ \int_{TF}^{\infty} (e^{-\gamma(t-TL-6)} - e^{-\gamma(t-TL-5)})Ve^{-\alpha -} dt$$

$$= V\frac{1}{\gamma + \rho}[e^{-\rho TL} - e^{-\gamma - \rho TL}] + V\frac{1}{\gamma + \rho}[e^{-\rho TL+6)} - e^{-\gamma - \rho(TL+6)}].$$

New Market with Imitation

Full replacement by farmers

In this case, we assume that the innovator introduces a new variety at T_L but that after a time lag imitators are able to introduce a competing variety at T_F; that is, with a time lag of $(T_F - T_L)$. We assume that imitators have a target market share of S_F and proceed towards the realization of this market share at the rate of 100μ per cent per year. With all farmers replacing seed every year, the innovator's quasi-rent becomes:

$$V_L = \int_{TL}^{\infty} (1 - e^{-\gamma(t-TL)})Ve^{-\alpha -} dt - \int_{TF}^{\infty} (S_F - S_F e^{-\mu(t-TF)})Ve^{-\alpha -} dt$$

$$= \frac{Ve^{-\rho TL}}{\rho}\left(\frac{1}{\rho} - \frac{1}{\gamma + \rho}\right) - S_F Ve^{-\rho TF}\left(\frac{1}{\rho} - \frac{1}{\gamma + \rho}\right).$$

No replacement by farmers

When farmers do not replace seed after the initial purchase of a new variety, the imitator can expect a share of only the new adopters' market every year after imitation. If the time lag for imitation is large, the imitator will have a very narrow market because almost all the adopters of the new variety would have made the initial purchase before the imitator entered the market. In such a situation the innovator's discounted quasi-rents will be:

$$V_L = \int_{TL}^{\infty} [(1 - e^{-\gamma(t-TL+1)}) - (1 - e^{-\gamma(t-TL)})]\ Ve^{-\alpha -} dt$$

$$-\int_{TF}^{\infty} S_F(1-e^{-\epsilon(TF-TL)-\mu(t-TF)})[(1-e^{-\gamma(t-TL+1)})-(1-e^{-\gamma(t-TL)})]\, Ve^{-\rho t}-dt.$$

The second expression shows the loss of quasi-rents to the innovator because the imitator takes up a share of the new adopters in each year following imitation. For simplicity in derivation, we can assume that the imitator reaches its target market share of new adopters instantaneously after imitation. (This amounts to assuming that $\epsilon=0$ and that the value of μ is sufficiently high for $(1-e^{-\epsilon(TF-TL)-\mu(t-TF)})$ to be approximately equal to 1.) The expression for V_L reduces to:

$$V_L=\int_{TL}^{\infty}(e^{-\gamma(t-TL)}-e^{-\gamma(t-TL+1)})Ve^{-\alpha}-dt-\int_{TF}^{\infty}S_F(e^{-\gamma(t-TL)}-e^{-\gamma(t-TL+1)})\, Ve^{-\alpha}-dt$$

$$=\frac{Ve^{-\rho TL}}{\gamma+\rho}(1-e^{-\gamma})-\frac{VS_F}{\gamma+\rho}(e^{-\gamma(TF-TL)-\rho TF})\,(1-e^{-\gamma}).$$

Similar expressions can also be derived when the imitator takes time to build up market share.

Replacement by a proportion of farmers every year

Let us assume that 20 per cent of the adopters of a variety replace seed every year. In this case the imitator can expect to sell not only to some of the new adopters but also to some of the old adopters who come to the market for replacement seed. Though the imitator will take time to reach its target market share, again for simplicity we assume that it reaches its target instantaneously. The innovator's discounted quasi-rent becomes:

$$V_L=\int_{TL}^{\infty}[(1-e^{-\gamma(t-TL+1)})-(1-e^{-\gamma(t-TL)})]\, Ve^{-\alpha}-dt+\int_{TL}^{\infty}(1-e^{-\gamma(t-TL)})\alpha Ve^{-\alpha}-dt$$

$$-\int_{TF}^{\infty}S_F[(1-e^{-\gamma(t-TL+1)})-(1-e^{-\gamma(t-TL)})]+\alpha(1-e^{-\gamma(t-TL)}]Ve^{-\alpha}-dt$$

$$=\frac{V}{\gamma+\rho}(e^{-\rho TL}-e^{-\gamma-\rho TL})+\alpha Ve^{-\rho TL}\left(\frac{1}{\rho}-\frac{1}{\gamma+\rho}\right)$$

$$-VS_Fe^{-\rho TL}\left[\frac{(1-\alpha)e^{-\gamma(TF-TL)}}{\gamma+\rho}-\frac{e^{-\gamma(TF-TL+1)}}{\gamma+\rho}+\frac{\alpha}{\rho}\right].$$

Replacement of seeds every *n* years

If farmers replace seed every five years, again the imitator can expect to sell to some of the new adopters every year in the years following imitation and also to some farmers who come to the market periodically for replacement seed. The innovator's discounted stream of quasi-rents can be derived as:

$$V_L = \int_{TL}^{\infty} [(1 - e^{-\gamma(t-TL+1)}) - (1 - e^{-\gamma(t-TL)})]\, Ve^{-\alpha} - dt$$

$$+ \int_{TL+5}^{\infty} (e^{-\gamma(t-TL-5)}) - e^{-\gamma(t-TL-4)})\, Ve^{-\alpha} - dt$$

$$- \int_{TF}^{\infty} S_F(1 - e^{-\epsilon(TF-TL)-\mu(T-TF)})\, [(1 - e^{-\gamma(t-TL+1)}) - (1 - e^{-\gamma(t-TL)})]\, Ve^{-\alpha} - dt$$

$$- \int_{TL+5}^{\infty} S_F(1 - e^{-\epsilon(TF-TL)-\mu(T-TF)})\, [(e^{-\gamma(t-TL-5)} - e^{-\gamma(t-TL-4)})\, Ve^{-\alpha} - dt.$$

Once again if we assume that the imitator reaches its target market imitation, the expression for V_L simplifies to:

$$V_L = V\frac{1}{\gamma+\rho}[e^{-\rho TL} - e^{-\gamma-\rho TL}] + V\frac{1}{\gamma+\rho}e^{-\rho TL+5)} - e^{-\gamma-\rho(TL+5)}$$

$$-\frac{VS_F e^{-\rho TL}}{\gamma+\rho}[e^{-\gamma(TF-TL)} - e^{-\gamma(TF-TL+1)}] - \frac{VS_F e^{-\rho(TL+5)}}{\gamma+\rho}[1 - e^{-\gamma}].$$

Impact of Seed Replacement Patterns and Imitation

The impact of farmer seed replacement behaviour on the innovator's discounted stream of quasi-rents can be seen with the help of an example in which the following values of different parameters are assumed:

$TL = 4$
$\gamma\ = 0.4$
$\alpha\ = 0.2$ (or in the alternative farmers replace seed every five years)
$V\ = 100$

As shown in Table 6.1, with an annual potential quasi-rent of 100, the discounted value of the innovator's quasi-rent, with no imitators and full replacement by farmers, is 396. But with no replacement by farmers the discounted value of quasi-rents falls drastically to just 9 per cent of the value with full replacement. When farmers replace seed every five years, the discounted value is 30 per cent of the value with full replacement, while it is 15 per cent when 20 per cent of farmers replace seed every year. The undiscounted stream of quasi-rents for different cases is shown in Figure 6A.1, while the discounted stream is shown in Figure 6A.2. The appropriation of returns in a situation with competitors is still lower for each of these cases.

If the discounted value with full replacement is taken to be the appropriation of returns from hybrids, it is clear that self-pollinated varieties yield only one-quarter of the returns from hybrids (assuming for the moment that development costs are similar). The low appropriability of returns from self-pollinated varieties acts as a disincentive to research. It is not surprising that, according to data for the USA hybrids attract almost four times the investment attracted by self-pollinated crops.

The analysis can be carried out similarly for the market rivalry situation, where the innovator's incremental rents will be even less. The analysis can be further modified for the case when there is a monopoly in the seed market to begin with. But the key point is that patterns of seed replacement behaviour have a drastic impact on appropriation of rents by the innovator.

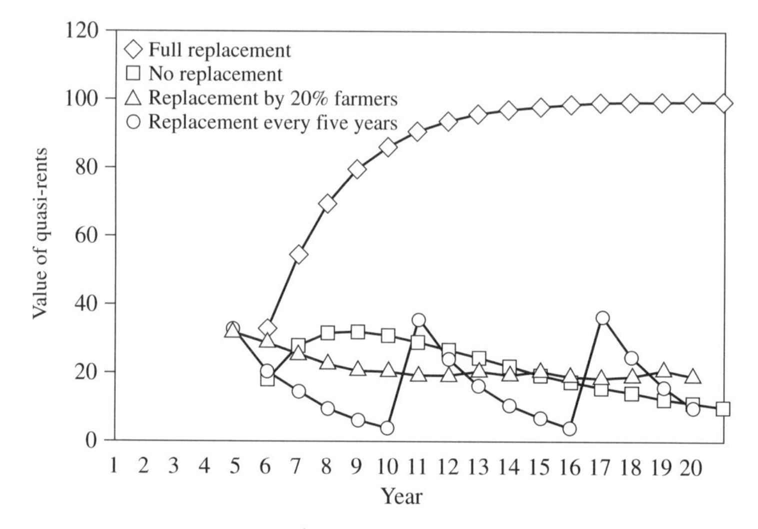

Figure 6A.1 Undiscounted quasi-rents for different seed replacement practices

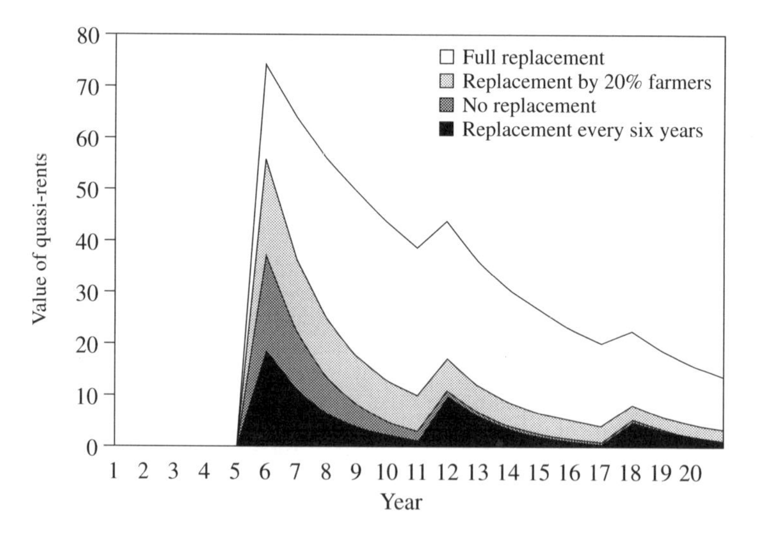

Figure 6A.2 Discounted value of quasi-rents under different seed replacement patterns

7. The impact of GURTs on developing countries: a preliminary assessment

Timothy Swanson and Timo Goeschl[1]

INTRODUCTION

Genetic use restriction technologies (GURTs)[2] represent a new technological advance in aid of the appropriation of the benefits generated by reason of plant breeding investments. Investments in R&D are always problematic, since the values of innovation are notoriously difficult to appropriate. In the case of plant breeding this problem is multiplied by the fact that the resulting product has in-built reproduction capabilities. This renders the need for future purchases unnecessary, and it makes it quick and inexpensive to become an instant competitor with the innovator in the use and supply of the innovation. Partly by reason of these characteristics of rapid diffusion, the agricultural R&D sector is currently a very diffuse and diverse enterprise, consisting of farmers, public sector enterprises, private multinationals and international organizations. GURTs have been devised to address the problem of appropriability, but they will also have substantial impacts on this overall R&D structure. The objective of this chapter is to attempt to disentangle some of these various impacts of GURTS, and especially to differentiate the likely effects on various developing countries.

The chapter proceeds by initially setting forth a discussion of the various factors that will determine the impacts of GURTs on various countries. These factors are then used to classify nearly 100 different developing countries by reference to the anticipated impact of the new technology. This exercise only points to the importance of a further study, as it is found that nearly half of the developing countries will have their benefits determined by the rate of diffusion that exists under the new technology. This leads to an examination of another system directed to the same purpose: the hybrid crop variety study that follows in the next chapter. The interrelationship between the appropriation system and the level of public spending is considered, before we conclude with a consideration of policy implications of the advent of GURTs.

DISCUSSION: THE FACTORS THAT WILL DETERMINE THE IMPACTS OF GURTs

The object of genetic use restriction technologies (GURTs) is to enhance the appropriability of the benefits from innovations in seed development. This object may be achieved in part through biotechnologies that engender future-generation seed sterility, making it necessary for farmers to repurchase seeds from the company at each planting season. This enhances appropriability because it means that farmers purchasing the seed once will not be able to compete with the seed company in the future supply of seed with those characteristics (for purposes of sale or own use).[3]

In the first instance there is no reason to believe that enhanced appropriability is harmful to the interests of farmers. This is because the *direct effect* of this technique will correspond only to the appropriation of the increased value of the innovations contained within the GURT seed. Farmers will continue to have the ability to purchase normally reproducing seed, only it will be without the innovations contained within the GURT seed. The availability of this seed will constrain the price at which GURT seed can be marketed, and it will mean that it will only be the added value of the innovation it contains that will cause it to be valued more highly. Therefore, at first sight, there is no reason to believe that farmers could be made worse off through the introduction of these technologies. These technologies merely add an option that did not exist previously.

However, there are significant *indirect effects* that may also result from the introduction of these technologies. For example, in the future, farmers refusing to purchase GURTs may be denied not only the single restricted use innovation (attached to a general use plant variety), they may also be denied the use of an entire series of past innovations that have never diffused into general agriculture. It is this potential impact on the diffusion of agricultural innovations that is the most problematic characteristic of this new technology. In order to see this, it is important to contrast how agricultural innovation diffused in the past with how it is likely to diffuse in the future.

In the past, innovations in plant varieties have diffused into general use within agriculture over time even if released as protected varieties, because of the capacity to undertake breeding activities making use of them.[4] Sometimes this breeding activity was undertaken by private individuals, but very often it occurred within public institutions. In fact publicly funded institutions (at both national and international levels) have put substantial funding and efforts into ensuring that recent innovations are diffused across the developing countries.[5] These public agricultural research institutions have done this by taking observed innovative characteristics and breeding them into locally used varieties.

Hence a big difference between GURTs and the previous appropriation systems is that GURTs capture the value of the innovative characteristics by maintaining control over the plant variety in which they are embedded. These distinctions between the 'innovative characteristics' and the 'plant variety templates' (on which they are embedded) are categorized as the 'software' and the 'hardware' components of modern plant breeding activities. To a significant extent in the past, even when the software has been developed by private breeders, it has been possible for much of it to diffuse quickly and inexpensively across the developing world by means of its incorporation within different hardware.

The movement toward GURTs may restrict the diffusion of recent innovations across developing countries. In effect the software would become hardware-specific, and it would be entirely up to the discretion and motivation of the innovator to diffuse its innovation to all of the various parts of the world on which it would confer benefits. In other words, the most significant (albeit indirect) impact of this change in technology would be the potential elimination of a currently diverse R&D sector (farmers, public sector, private sector) and its replacement by a fairly homogeneous and highly concentrated private sector. It is possible, then, that the rate and extent of diffusion of future innovations in agriculture would occur only under the exclusive control and direction of the originator of the initial innovation.

This is a problem for two reasons.[6] First, with sufficient time and a significant number of innovations, the other currently existing suppliers of plant varieties (private and public) might be rendered commercially obsolete. This would be the case if the alternative suppliers could only acquire the characteristics at high prices, and thus were only able to supply inferior substitute varieties. Then the farmer might face a small number of suppliers of viable seed, and consequentially much increased prices for GURT seed varieties.[7] Secondly, if this were the case, then the private sector might be able effectively to eliminate the public sector from all breeding activities on account of the need for licences and the restrictiveness of material exchanges. This might have deleterious consequences for those countries that are most highly dependent on public investment for their plant breeding needs.

Another perspective on this part of the problem is to note that the commercial sector may not have sufficient private incentives to diffuse their software widely, that is, across a diverse enough array of plant varieties.[8] Then the innovations would be aimed only at those markets where there was adequate demand, while general diffusion would be disallowed in order to protect those markets. The farmers on the fringes would be faced with farming with the innovative characteristics embedded in poorly performing varieties, or farming with the best local varieties but without the innovative characteristics.

Therefore the move to GURTs as an appropriation system is important primarily for the indirect effects that it might have on the entire system of R&D currently existing within agriculture. And these indirect effects are important because they might make a tremendous difference in the rate of diffusion of innovations to particular developing countries. In the next sections we first attempt to distinguish between the various categories of countries, and then we attempt to assess the extent to which the rate and direction of innovation will be slowed in those countries that are most affected.

CATEGORIES OF COUNTRIES

The above discussion indicates that there are a few key factors that will determine the impact of these new technologies on the various developing countries. The first factor is the capability of the developing country to undertake its own biotechnology. If it is able to do this, there is little change in the rate of technological diffusion with the introduction of GURTs, as these countries will be able to 'reverse engineer' GURT varieties as easily as they could any other variety.[9] For these countries (the biotechnology-capable) the impacts of GURTs are primarily positive. So the first important question for ascertaining the impact of this new technology is: *does the subject country possess actual or incipient biotechnological capabilities?*

For countries without biotechnological capabilities, the important question concerns the impacts of GURTs on the rate of diffusion of innovations within their agriculture. In order to address this issue, we turn to an analogue from the agricultural industry. Approximately 50 years ago the agricultural industry experienced its first technological revolution with the introduction of modern hybridized varieties (into sexually reproducing crops such as maize). To a large extent the advent of GURTs simply extends the effects of these forms of technologies to asexually reproducing crops (such as wheat and rice). The second important question for ascertaining the impact of GURTs is: *does the subject country have a significant investment in crops that are amenable to GURTs (such as wheat and rice)?*

The third important question will concern the extent to which the biotech-capable countries will include another country within their research strategies. *How quickly will the benefits from the technology diffuse to the non-biotech-capable country?* Again, the experience with the hybridized modern varieties is instructive. We are able to examine how quickly and how extensively an individual country has benefited from innovations in maize breeding, in order to see the extent to which that country fits within the set of countries that will benefit from genetic use restriction technologies more generally.

In the remainder of this chapter we attempt to assess qualitatively the impacts of GURTs on various developing countries by reference to these specific factors. The first two questions are addressed by reference to various indicators of technological capability. The last question is addressed in a case study on the impacts of hybrid variety-based technologies.

IMPACT GROUPS AND GURTs

There are three basic dimensions along which the economic impacts at the country level will be assessed. We will assume that GURTs will have the effect of increasing appropriability of the value (or rents) from plant breeding activities, given that the country has biotechnological (BT) capability. We will further assume that GURTs will have positive impacts on the productivity of agriculture in those countries with (a) BT capability and (b) significant land area dedicated to GURT crop varieties. Finally, we will assume that when a country is not BT-capable, the key to the impacts of GURT technologies will be the rate and extent of the diffusion of innovation to that country; that is, to what extent do the needs of the given country factor into the plans and objectives of the BT-based plant breeding sector?

These assumptions lead to the categorization of the developing countries into five distinct groups. Table 7.1 gives a summary of the classifications. *Group A* comprises those countries with existent BT capabilities. These are the developing countries that will produce GURT-based crops themselves. *Group B* includes those with an incipient biotech sector. This may be the result of either an already existent but immature sector, or a good prospect for the development of GURT capacity through foreign investment and ease of regulation. In other words, these are the countries with good potential to catch up to group A.

In *group C* we classify countries likely to convert their agriculture to GURT-based systems and thus reap the productivity benefits from that technology although they will be unable to develop domestic GURT capacity for the time being. They are countries with a built-in tendency towards

GURT crops, and a moderately high rate of diffusion of innovations from other countries' plant breeding sectors. *Group D* features those countries that are currently highly dependent on public R&D spending and may lack the liquidity to adopt risky new technology. This means that these countries are in danger of suffering disadvantages from GURTs, particularly by virtue of a slowdown in the rate of agricultural productivity growth. They are 'slow diffusion' countries, in that innovations from other countries' plant breeding sectors do not necessarily confer productivity benefits in these countries.

Finally, *group E* consists of those countries with a small amount of land in crops likely to be the target of GURTs. Although these countries are not likely to benefit from this biotechnology in any way, they are also unlikely to find themselves in a worse position than before. These are countries that will face a small loss in R&D rents, but whose level of productivity growth – already low – will be unaffected by GURTs.

Table 7.1 Classification of impact groups with respect to GURTs

Group	Likely impacts
(A) BT capable	Immediate increase in appropriation of R&D rents Immediate increase in R&D Immediate increase in rate of agricultural productivity growth
(B) BT incipient	Medium-term increase in appropriation of R&D rents Medium-term increase in R&D Immediate increase in rate of agricultural productivity growth
(C) Moderate diffusion	Decrease in appropriation of R&D rents Increase in rate of agricultural productivity growth
(D) Slow diffusion	Decrease in appropriation of R&D rents Productivity impact dependent on rate of diffusion
(E) No diffusion	Appropriation of R&D rents unaffected Productivity growth unaffected

CRITERIA FOR CLASSIFICATION

As indicated above, the primary indicators of GURT impacts are (1) existence of BT capabilities, (2) potential to develop BT capabilities, (3) rate and extent of diffusion of innovations from the BT capable, and (4) the impact of GURTs on the current crop portfolio. In this section we select certain criteria to proxy for the above indicators. These are listed below in Table 7.2.

Table 7.2 Criteria for classification of impact groups

Criterion	Explanation
Indicator 1: biotech capacity or stage of development	This criterion captures the current stage of the biotechnological capability of a developing country. 'Capable' means that a country is currently able to produce genetically modified organisms. The other terms refer to the stage of development of biotech capacity. 'Advanced' means that it is within 5 years of being able to do so, 'nascent' means that a country is likely to become capable within 10 years. 'Preparatory' means that a country has taken first steps and provided public funding for the establishment of biotech capability. The countries named have been identified in an ISNAR survey report (Komen and Persley, 1993).
Share of world area of transgenic crops	These data show how much of the world's transgenic crop is grown in a particular country, thus indicating a country's agrotechnological stage and experience with GM crops. The data were compiled by ISAAA in 1998 (James, 1998).
Indicator 2: FDI ease of investment	This criterion reports a country's ranking in terms of openness to foreign direct investment (FDI). It is therefore a measure of a country's likelihood of benefiting from the transfer of biotechnological knowledge from developed countries by direct investment in human and physical capital. Countries are ranked from 1 (very open) to 5 (closed to FDI). The ranking was conducted by the Heritage Foundation in 1998 (Johnson and Holmes, 1998).
REG regulatory environment	This criterion measures a country's ranking in terms of costliness and impediments to economic freedom by virtue of governmental regulation. Thus it is a measure of how rapidly a country can benefit from technological progress through implementation. Countries are ranked from 1 (low level of regulation) to

Table 7.2 (continued)

Criterion	Explanation
	5 (very high level of regulation). The ranking was conducted by the Heritage Foundation in 1998 (Johnson and Holmes, 1998).
Indicator 3: yield gap	These data show the gap between the country's yield in maize and the average maize yield for developed countries as a percentage of the developed country mean. This is a rough indicator of a country's current ability to capture productivity rents from a crop for which the most productive varieties are technologically protected. Countries with a high gap have problems at present in fully capitalising on the best technology available. This data are compiled on the basis of the most recent FAO database entries for 1998/9.
Indicator 4: GURT potential	These data show what share of a country's arable land not planted in crops with hybrid potential is planted in crops likely to be the target of GURTs (wheat, rice, cotton, soybean, barley). This indicates the potential for introducing GURTs into this country under the present conditions. Countries that have either exhausted their arable land in the cultivation of crops that are available as hybrids or grow an exotic portfolio of crops achieve a low score in this category. These data are compiled on the basis of the most recent FAO database entries for 1998/9.

We use the measured stage of development of the biotechnology sector previously, as the proxy for existence of BT capacity. Most of these countries should be listed in the first rank of the biotechnologically capable, by reason of their head start in the area. The other proxy we use for this variable is the actual area of the world's transgenic crops. To measure the potential for biotechnological development, we use an index measuring the ease of foreign investment and another measuring the stringency of regulation. It is assumed that the transfer of high technology, such as BT, will require substantial foreign investments and a relatively light touch in regulation.

To measure the extent to which BT innovations diffuse to the subject country, we use a measure derived from the case of maize hybrid varieties. This measure indicates the extent to which the given country is operating at the technological frontier in maize production. To the extent that it falls short of the technological frontier, this must be to some extent attributable to the failure of innovations to be tailored or diffused to the subject country's characteristics. Finally, to measure the potential (non-)impact of GURTs, we use a measure of the extent to which arable land is currently invested in crops that are likely to be early subjects of GURTs. To the extent that a country has not invested in such crops, it is likely that the change in technology will have little, if any, impact on that country.

CLASSIFICATION OF DEVELOPING COUNTRIES

Table 7.3 classifies 98 developing countries in accordance with the indicators and criteria set forth above. *Group A* is the class of countries for which GURTs are likely to have an immediate and positive impact. *Group B* is the class of countries for which GURTs are likely to have a positive impact in the near term (once biotechnological capacities are advanced). *Group C* is the class of countries for which GURTs are likely to have an uncertain impact because of countervailing effects. They are unlikely to benefit from their own biotechnological capacities, but they have indicated that they do benefit from relatively rapid and extensive diffusion of innovations from other countries. *Group D* is the class of countries for which GURTs are likely to have an uncertain impact, highly dependent on the extent to which innovations diffuse from the capable to the non-capable countries. *Group E* is the class of countries for which GURTs are likely to have little impact, positive or negative.[10]

THE DIFFUSION PROBLEM

Table 7.3 classifies 40 out of the 95 developing countries assessed as belonging to group D. This makes it by far the most prominent group among the five different classes. For reasons explained below, this is also the group for which it is most difficult to assess the net impact of GURTs.

Group D consists of countries that are not likely to command biotechnologies and their commercial application for the foreseeable future. This means that they will certainly be adversely affected by the shift in the share of R&D rents from farmers and consumers to plant breeders. With no significant R&D contributions that could be protected by the use of GURTs, the R&D balance will move against these countries. On the other

Table 7.3 Developing countries grouped according to GURT impacts (in accord with the indicators and measures discussed below)

Country	Biotech capacity/ stage of development	FDI index: ease of investment	REG: regula-tory envmt	% of world total of transgenic crop area	GURT crop potential as share of arable land (%)	Yield gap to tech. frontier average in maize (%)
Group A (positive impact – now)						
Argentina	Capable	2	2	15	64	−29
Brazil	Capable	3	3	0	45	−61
Chile	Capable	2	2	0	23	25
China	Capable	3	4	14	77	−27
India	Capable	4	4	0	60	−76
South Africa	Capable	2	2	1	8	−66
Thailand	Advanced	2	3	0	68	−51
Turkey	Advanced	2	3	0	51	−45
Group B (positive impact – near term)						
Colombia	Preparatory	2	3	0	46	−75
Egypt	Nascent	3	4	0	100	−2
Indonesia	Advanced	2	4	0	87	−63
Jordan	n/a	2	3	0	44	−11
Kenya	Preparatory	3	4	0	9	−77
Malaysia	Advanced	3	2	0	36	−75
Mexico	Capable	2	4	1	9	−68
Paraguay	n/a	1	4	0	87	−66
Philippines	Nascent	3	4	0	100	−77
Zimbabwe	Preparatory	4	4	0	31	−85
Group C (uncertain impact – moderate diffusion)						
Algeria	n/a	3	3	0	41	−69
Bolivia	n/a	2	4	0	53	−54
Iran	n/a	5	4	0	44	−14
North Korea	n/a	5	5	0	86	−49
South Korea	n/a	2	3	0	72	−42
Laos	n/a	5	5	0	100	−65
Surinam	n/a	3	4	0	95	−67
Uruguay	n/a	2	3	0	42	−63
Vietnam	n/a	4	5	0	100	−65

Table 7.3 (continued)

Country	Biotech capacity/ stage of development	FDI index: ease of investment	REG: regula-tory envmt	% of world total of transgenic crop area	GURT crop potential as share of arable land (%)	Yield gap to tech. frontier average in maize (%)
Group D (uncertain impact – slow diffusion)						
Afghanistan	n/a	n/a	n/a	0	34	−83
Bangladesh	n/a	3	5	0	100	−85
Benin	n/a	3	3	0	76	−82
Bhutan	n/a	n/a	n/a	0	45	−88
Burkina-Faso	n/a	2	4	0	22	−80
Cambodia	n/a	3	4	0	55	−83
Chad	n/a	4	4	0	18	−81
Costa Rica	n/a	2	3	0	39	−74
Ecuador	n/a	2	4	0	50	−83
Eritrea	n/a	n/a	n/a	0	61	−89
Ethiopia	n/a	4	4	0	24	−77
Guinea	n/a	3	4	0	67	−85
Guinea-Bissau	n/a	4	5	0	25	−86
Guyana	n/a	3	4	0	30	−83
Haiti	n/a	4	5	0	32	−89
Iraq	n/a	5	4	0	58	−77
Ivory Coast	n/a	3	4	0	47	−88
Kuwait	n/a	4	2	0	100	?
Lebanon	n/a	3	3	0	20	−66
Lesotho	n/a	3	4	0	92	−87
Madagascar	n/a	4	3	0	52	−88
Malawi	n/a	3	4	0	57	−75
Mali	n/a	2	3	0	22	−76
Mongolia	n/a	3	4	0	23	?
Morocco	n/a	2	3	0	47	−94
Mozambique	n/a	4	5	0	31	−87
Myanmar	n/a	4	5	0	65	−75
Nepal	n/a	4	4	0	100	−76
Nigeria	n/a	2	4	0	18	−81
Pakistan	n/a	2	4	0	70	−81
Peru	n/a	2	3	0	20	−70
Qatar	n/a	3	4	0	12	?
Sierra Leone	n/a	3	4	0	71	−87
Sri Lanka	n/a	3	3	0	99	−84
Swaziland	n/a	2	3	0	31	−73
Syria	n/a	4	4	0	66	−77

Table 7.3 (continued)

Country	Biotech capacity/ stage of development	FDI index: ease of investment	REG: regulatory envmt	% of world total of transgenic crop area	GURT crop potential as share of arable land (%)	Yield gap to tech. frontier average in maize (%)
Tanzania	n/a	3	4	0	100	−81
Tunisia	n/a	2	3	0	46	?
Venezuela	n/a	3	3	0	10	−63
Yemen	n/a	4	4	0	20	−79
Group E (low impact)						
Angola	n/a	4	5	0	2	−89
Belize	n/a	2	3	0	12	−70
Botswana	n/a	3	3	0	0	−97
Burundi	n/a	4	4	0	6	−84
Cameroon	n/a	3	4	0	4	−79
Congo (Z)	n/a	5	4	0	12	−88
Cuba	n/a	5	5	0	4	−82
Dom. Rep.	n/a	3	4	0	13	−82
El Salvador	n/a	1	3	0	15	−72
Gabon	n/a	2	3	0	1	−75
Gambia	n/a	4	4	0	11	−80
Ghana	n/a	3	4	0	10	−79
Guatemala	n/a	3	4	0	5	−72
Honduras	n/a	3	4	0	1	−86
Libya	n/a	5	5	0	13	−84
Mauritania	n/a	3	4	0	7	−91
Namibia	n/a	2	3	0	0	−89
Nicaragua	n/a	2	4	0	4	−84
Niger	n/a	4	4	0	1	−83
Oman	n/a	4	3	0	4	?
Panama	n/a	2	3	0	13	−82
PNG	n/a	3	4	0	1	?
Rwanda	n/a	4	5	0	2	−85
Saudi Arabia	n/a	4	3	0	11	−72
Senegal	n/a	3	4	0	6	−86
Somalia	n/a	4	5	0	4	−91
Sudan	n/a	4	4	0	2	−96
Togo	n/a	4	5	0	12	−87
Uganda	n/a	2	2	0	8	−82
UAE	n/a	4	2	0	1	?
Zambia	n/a	2	4	0	2	−78

hand, with reasonable shares of land under potential GURT crops, improvements in the rate of agricultural productivity growth are possible.

The problem for the countries in group D lies in their current yield gap in maize. This is the third defining criterion of countries in group D. It identifies countries that lag far behind the yield development in another crop already protected by technological means, maize, and that are therefore likely to face similar problems when technological protection becomes available for other crops. This unsatisfactory track record in catching up to productivity developments in maize shifts the focus to the problem of the diffusion of crop improvement across countries. Many countries in group D rely heavily on the 'trickle-down' process of productivity gains that is mediated through public plant breeding institutions and local farmer breeding (Conway and Thoenissen, 2000). The important issue becomes the extent to which the maize experience is generalizable to GURTs. To answer such questions it is necessary to turn to a case study on the impact of the introduction of hybridization technologies on the character of the R&D sector (See hybrid variety study in Chapter 8).

PREVIEW OF THE RESULTS FROM THE HYBRID VARIETY STUDY

The key results from the study in the next chapter are indicative of severe difficulties for the countries classified in Group D. These results are as follows: (a) the new appropriation technology stimulates private R&D investment; (b) the new technology 'crowds out' public sector plant breeding in the area; (c) the combination of increased private R&D investment and reduced public sector investment results in a widening productivity gap between the developed and the developing countries.

In short, the availability of hybrid varieties in maize increased private plant breeding activities but reduced involvement in maize plant breeding by the public sector. This outcome indicates that the integration of the 'software' and 'hardware' of plant breeding makes the private sector more profitable, while rendering the public sector infeasible. The public sector no longer offers a generally competitive alternative to the private sector's plant varieties, and so it leaves the field to the private sector.

The outcome is a widening gap between those countries whose plant breeding is on the technological frontier and those whose breeding lags behind. Clearly, the private sector is not placing the same amount of efforts into diffusing its innovations throughout the developing countries as is the private/public mixed R&D sector for those crops unable to be hybridized. Just as clearly, the public sector is not able to continue to function in these markets, probably by reason of the restricted flow of materials and characteristics. The replacement of the mixed system of R&D with the private sector version has resulted in systematic widening of productivity gaps between rich and poor countries.

THE PUBLIC SECTOR CONTEXT: PUBLIC SECTOR SPENDING AND TECHNOLOGICAL CHANGE

The critical factor for the countries in group D is the rate of diffusion that will exist under GURTs. The hybrid variety study in the next chapter indicates that there is good reason to be concerned that the rate of diffusion will slow with GURTs, as the flow of plant materials and the level of public funding is restricted.

It may be possible to avoid repeating this experience with the advent of GURTs. The advance of the private hybrid sector has taken place against the background of a policy of declining emphasis on public sector agricultural R&D. The growth in spending on agricultural R&D has been slowing down in developed countries, thus giving rise to a slowing down in the rate of diffusion. Global agricultural R&D spending disaggregated by developed and developing countries is presented in Table 7.4. The combination of increasingly complex technologies, increasingly restricted (and costly) material exchange and declining growth in public R&D investments makes it impossible for public sector plant breeders to maintain the breadth of objectives that existed previously. It is a combination of three factors – technological advance, material exchange restrictions and public investment levels – that will determine the overall impact on developing countries dependent on diffusion.

THE IMPACT OF GURTs: POLICY CHOICES

Since the impact of these appropriation technologies is likely to be dependent on other public policies, it is important to consider the interaction between these policies and technologies. It is possible that the impacts of GURTs might differ significantly from those of the hybrids, if the public sector combines them with different funding and management policies.

Table 7.5 displays the rates of R&D diffusion for three different technology management systems and under two different assumptions concerning the level of public spending. For example, the table indicates that the rate of diffusion of innovations in hybrid varieties is slow with low levels of public funding (as the maize case study indicates) but is likely to increase

Table 7.4 Real agricultural R&D spending in developed and developing countries, 1971–91

	Expenditures (millions of 1985 international dollars)		
	1971	1981	1991
Developing countries (131)[a]	2 984	5 503	8 009
Sub-Saharan Africa (44)[a]	699	927	968
China	457	939	1 494
Asia and Pacific (excl. China) (28)[a]	861	1 922	3 502
Latin America and Caribbean (38)[a]	507	981	944
West Asia and North Africa (20)[a]	459	733	1 100
Developed countries (22)[a]	4 298	5 713	6 941
Total (153)[a]	7 282	11 217	14 951
	Average annual growth rates (per cent)		
	1971–81	1981–91	1971–91
Developing countries	6.4	3.9	5.1
Sub-Saharan Africa	2.5	0.8	1.6
China	7.7	4.7	6.3
Asia and Paciic (excl. China)	8.7	6.3	7.3
Latin America and Caribbean	7.0	−0.5	2.7
West Asia and North Africa	4.3	4.1	4.8
Developed countries	2.7	1.7	2.3
Total	4.3	2.9	3.6

Note:
[a] Figures in parentheses indicate the number of countries included in the respective totals.

Source: Alston *et al.* (1998).

Table 7.5 Rate of R&D diffusion to countries dependent on funding (under different technology management systems and levels of public R&D spending)

Appropriation system		Public funding: Low	Public funding: High
Current IPR	Hybrid varieties	Slow	Moderate
	Non-hybrid varieties	Moderate	Fast
IPR + GURTs		Slow	Moderate
GURT only		Slow/moderate	Fast

significantly if public sector funding is increased. On the other hand, the diffusion of innovations for non-hybrid varieties is currently much more rapid and extensive than it is for hybrid varieties (but also dependent on public sector expenditures). Hence Table 7.5 indicates that agricultural R&D currently registers very different impacts on developing countries, depending on whether the variety is a hybrid one or not (cf. Srivastava *et al.*, 1996).

The critical issue for the impact of GURTs is whether (given the combination of management and spending policies applied) they will more closely follow the current example of the hybrid or the non-hybrid sector. What are the different appropriation systems and what are their potential implications for diffusion? Table 7.6 explains the two different forms of appropriation thought possible for GURTs, in contrast to the current system of IPR protection that forms the baseline scenario.

Table 7.6 Baseline and other scenarios for appropriation systems

System	Description
Current IPR-based system (baseline)	This is the current system, where the release of GURTs is delayed indefinitely. It features one sector where legal protection through IPRs must be sought to protect R&D investment while it offers technological protection for a small set of outbreeding crops through hybridization. Under the current system, IPR remains the sole source of protection for R&D inputs into crop improvement in wheat, rice, cotton and other staple crops.
Combined use of GURTs and IPRs	This scenario regarding the legal implementation of GURTs can be considered as probable, as it does not require any changes in the existing legal structure. The situation would closely resemble the one that presents itself currently with combined legal protection in the form of *plant variety protection* (PVP) and technological protection through the use of *hybrid cultivars*. The introduction of GURTs would extend the feasibility of these combined regimes (beyond outbreeding crops). It is likely to have the effect of restricting the vast majority of agricultural R&D to the private sector, and of slowing the diffusion of innovation to developing countries in Group D.

Table 7.6 (continued)

System	Description
GURTs only – public sector policies directed to diffusion	It is possible for a combination of public sector policies to make the advent of GURTs a win–win situation. This would be the case if public policies were directed to the purpose of speeding the diffusion of private sector innovations (protected only by GURTs) throughout general agriculture. This would occur by means of enhanced public investment in R&D for the express purpose of reverse engineering the GURT-protected characteristics, with the object of translocating them into the local varieties for developing countries. GURT-based protection probably would be adequate to maintain appropriability of a new plant variety for about three years, and so it would afford some protection to the innovators. The removal of other (IPR-based) constraints on diffusion would aid the rapid extension of these innovations to developing countries.

CONCLUSION

GURTs are positioned to become an additional technological solution to the problem of rent appropriation. They represent the capacity to extend the solution concept of hybridization to those crop varieties that are not outbreeding. This suggests that the rates of diffusion for hybrids (in Table 7.5) will be extended to those crops that are currently protected only through IPRs. This implies a significant decline in the rate of diffusion of innovations across agriculture. In turn, this suggests a problem for those developing countries identified previously as dependent on the public sector for the diffusion of agricultural innovations.

The only real alternative is to meet this technological advance with public policies addressed to its deficiencies. These will be public policies that are directed to the purpose of aiding the diffusion of these innovations from the private sector to the public sector, and then on to those developing countries with little biotechnological capability. In short, the public sector must manage these deficiencies in diffusion with funding directed to the reverse engineering of GURT-protected innovations, and the diffusion of these characteristics into general agriculture. It must also make clear that GURTs will function well as an alternative to IPR-based protection systems, and not as a complement to them.

NOTES

1. This chapter is a by-product of a project financed by the UK Department for International Development on 'The Impacts of GURTs on the Urban and Rural Poor in Developing Countries'. We are extremely grateful for the contributions from the others involved in this project, including Philippe Sands, Terry Fisher, Colin Thirtle, Michael Lipton, Ed Barbier, Jonathon Jones and Philip Dale. We would also like to thank Calestous Juma for many helpful comments.
2. GURTs may come as either variety-based or trait-based. Since the described technologies are predominantly variety-based, this chapter will focus only on these. The authors do not believe that the problems indicated herein would apply to trait-based GURTs.
3. An alternative method of achieving this same object is the adoption of laws restricting such resale or reuse of commercially acquired seed. This method has been implemented in those countries that have adopted so-called 'plant breeders' rights' (PBRs) and/or seed patents, and enforced them strictly against their citizens. To a large extent GURTs should be seen as a technologically supplied alternative to these systems, with the important difference that individual countries do not have the option under GURTs to elect adoption of the system or determine the degree of enforcement. In this chapter we examine the impact of switching from such IPR-based R&D systems to GURT-based systems.
4. There are several important distinctions between PBR systems and GURT-based systems. First, PBR systems are limited in duration, while GURTs are not. Second, PBR systems often contain an 'own use' exemption for farmers that enable their own breeding activities. Thirdly, and most importantly, PBR systems merely disallow the marketing of the same plant variety in competition with its innovator; they do not disallow breeding activities making use of the new plant variety (for example, to translocate its innovative characteristics to other plant varieties).
5. The domestic system used for accomplishing this purpose is known as the national agricultural research centres (NARCs) while the international system devised in part to achieve the same purpose is known as the Consultative Group for International Agricultural Research (CGIAR system).
6. Again, it is important to emphasize that there is no problem in the first instance about the appropriation of the returns from innovation – this will simply enhance the prospects for effective investments in plant breeding activities. The problems that must be considered are second-order ones, namely the impact of wholly vesting agricultural R&D within the private sector and within a concentrated industry.
7. This would be the result of the refusal to licence innovations at reasonable prices to potential competitors, and the maintenance of very low prices until they were removed from the market. It is essentially the sort of conduct with which Microsoft has been charged.
8. For example, one well-known problem in the economics of market failure is the difficulty imposed by minimum efficient scales of operation. This can result in concentrating only on the median consumer, or in this case the typical agricultural producer, and the failure to address the diversity of needs within a diverse consumer base.
9. The GURT system has little effect on biotechnology-capable countries, precisely because they will be able to use biotechnology to unravel and relocate the innovative characteristic. In this case the GURT merely provides a short-term advantage to the innovative breeder, possibly a head start of only two or three years in the marketing of the charac-

teristic. The problem lies more in those countries with both little of their own biotechnological capacity and little investment by others interested in diffusing innovative characteristics into their local varieties.

10. It should be noted that the grouping used here is intended only for illustrative purposes. Clearly, there may be better proxies and measures of the indicators we have discussed, and these would provide for different groupings of the developing countries. The sole purpose of Table 7.3 is to indicate, firstly, that the impacts of GURTs will be non-uniform and highly variable and, secondly, that the variability of the impacts will nevertheless be systematic and reasonably predictable.

REFERENCES

Alston, J.M. and R.J. Venner (1998), 'The Effects of U.S. Plant Variety Protection Act on Wheat Genetic Improvement', paper presented at the symposium on 'Intellectual Property Rights and Agricultural Research Impact' sponsored by NC 208 and CIMMYT Economics Program, 5–7 March, El Batan, Mexico.

Alston, J.M., P.G. Pardey and Vincent H. Smith (1998), 'Financing Agricultural R&D in Rich Countries: What's Happening and Why?', *The Australian Journal of Agricultural and Resource Economics*, 42(1), 51–82.

Butler, L.J. (1996), 'Plant Breeders' Rights in the U.S.: Update of a 1983 Study', in J. Van Wijk and Walter Jaffe (eds), *Proceedings of a seminar on 'The Impact of Plant Breeders' Rights in Developing Countries' held at Santa Fe Bogota, Colombia, 7–8 March 1995*; Amsterdam: University of Amsterdam, pp. 17–33.

Butler, L.J. and B.W. Marion (1985), *The Impact of Patent Protection on the U.S. Seed Industry and Public Plant Breeding*, Food Systems Research Group Monograph 16, University of Wisconsin Madison, Madison, WI.

Capalbo, S.M. and J.M. Antle (1989), 'Incorporating Social Costs in the Returns to Agricultural Research', *American Journal of Agricultural Economics*, 71, 458–63.

CIMMYT (1999), *A Sampling of Impacts 1999. New Global and Regional Studies*, Mexico: CIMMYT.

Conway, G. and G. Thoenissen (2000), 'Biotechnology, Food and Nutrition', *Nature*, Millennium Supplement.

Crouch, M. (1998), 'How Terminator Terminates', rev. ed. Edmonds Institute Occasional Papers, Edmonds Institute, Washington.

Evenson, R.E. (1989), 'Spillover Benefits of Agricultural Research: Evidence from US Experience', *American Journal of Agricultural Economics*, 71, 447–52.

Falck-Zepeda, J.B. and Greg Traxler (1998), 'Rent Creation and Distribution from Transgenic Cotton in the U.S.', paper prepared for the symposium 'Intellectual Property Rights and Agricultural Research Impacts', sponsored by NC-208 and CIMMYT Economics Program, 5–7 March, CIMMYT Headquarters, El Batan, Mexico.

Frey, K.J. (1996), 'National Plant Breeding Study – I: Human and Financial Resource Devoted to Plant Breeding Research and Development in the United States in 1994', Special Report 98, Iowa Agriculture and Home Economics Experiment Station, Iowa State University, Iowa.

Fuglie, K., Nicole Ballenger, Kelly Day (1996), 'Agricultural Research and Development: Public and Private Investments Under Alternative Markets and Institutions', AER-735, Economic Research Service, United States Department of Agriculture.

Gupta, A. (1999), *Biosafety in an International Context*, Cambridge, MA: Harvard University Press.

Huffman, W.E. and R.E. Evenson (1993), *Science for Agriculture: A Long-Term Perspective*, Ames, IA: Iowa State University Press.

Jaffe, W. and J. Van Wijk (1995), 'The Impact of Plant Breeders' Rights in Developing Countries: Debate ad Experience in Argentina, Chile, Colombia, Mexico and Uruguay', Inter-American Institute for Co-operation in Agriculture and University of Amsterdam, Amsterdam.

James, C. (1998), 'Global Review of Commercialized Transgenic Crops: 1998', ISAAA Brief no. 8, Ithaca, NY.

Jarvis, D. and T. Hodgkin (eds) (1998), *Strengthening the Scientific Basis of In Situ Conservation*, Rome: IPGRI/FAO.

Jefferson, R.A. with D. Byth, C. Correa, G. Otero and C. Qualset (1999), 'Genetic Use Restriction Technologies. Technical Assessment of the Set of New Technologies which Sterilize or Reduce the Agronomic Value of Second Generation Seed as Exemplified by U.S. Patent no. 5,723,765 and WO 94/03619', expert paper prepared for the Secretariat of the Convention for Biological Diversity, Subsidiary Body on Scientific, Technical and Technological Advice, Montreal, Canada.

Johnson, B. and K. Holmes (1998), *The 1999 Index of Economic Freedom*, New York: Dow Jones and Co.

Kalton, R.R. and P.A. Richardson (1983), 'Private Sector Plant Breeding Programmes: A Major Thrust in U.S. Agriculture', *Diversity*, 1(3), 16–18.

Kalton, R.R., P.A. Richardson and N.M. Frey (1989), 'Inputs in Private Sector Plant Breeding and Biotechnology Research Programs in the United States', *Diversity*, 5(4), 22–5.

Komen, J. and G.J. Persley (1993), Agricultural Biotechnology in Developing Countries: A Cross-Country Review', ISNAR *Research Report no. 2*, The Hague: ISNAR.

Leisinger, K. (1995), 'Sociopolitical Effects of New Biotechnologies in Developing Countries', 2020 Vision Discussion Paper 2, IFRPI, Washington, DC.

Leisinger, K. (1997), 'Ethical and Ecological Aspects of Industrial Property Rights in the Context of Genetic Engineering', paper presented at Interlaken Conference.

Leonard, H.J with M. Yudelman, J.D. Stryker, J.O. Bowder, A.J. de Boer, T. Campbell and A. Jolly (1989), *Environment and the Poor: Development Strategies for a Common Agenda*, New Brunswick: Transaction Books.

Lesser, W. (1990), 'Sector Issue II: Seeds and Plants', in W.E. Siebeck (ed.), *Strengthening Intellectual Property Rights in Developing Countries: A Survey of Literature*, Madison, WI: Wisconsin University Press, pp. 59–68.

Perrin, R.K., K.A. Kunnings and L.A. Ihnen (1983), 'Some Effects of the U.S. Plant Variety Protection Act of 1970', Economics Research Report no. 46, Department of Economics and Business, North Carolina State University.

Pray, C.E., M.K. Knudson and L. Masse (1993), 'Impact of Changing Intellectual Property Rights on U.S. Plant Breeding R&D', mimeo, Department of Agricultural Economics, Rutgers University, New Brunswick.

Pray, C.E., S. Ribeiro, Rolf A.E. Mueller (1991), 'Private Research and Public Benefit – The Private Seed Industry for Sorghum and Pearl-Millet in India', *Research Policy*, 20(4), 315–24.

Qaim, M. and J. von Braun (1998), 'Crop Biotechnology in Developing Countries:

A Conceptual Framework for Ex Ante Economic Analysis', ZEF Discussion Papers on Development Policy, no. 3. ZEF, Bonn.

Reilly, J. and T. Phipps (1988), 'Technology, Natural Resources and Commodity Trade', in John D. Sutton (ed.), *Agricultural Trade and Natural Resources: Discovering Critical Linkages*, Boulder, CO: Lynne Rienner Publishers, pp. 224–35.

Rejesus, R., M. van Ginkel and M. Smale (1998), 'Wheat Breeders' Perspectives on Genetic Diversity and Germplasm Use. Findings from an International Survey', WPSR no. 40, CIMMYT.

Schmidt, J.W. (1984), 'Genetic Contributions to Yield Gain in Wheat', in W.R. Fehr (ed.), *Genetic Contributions to Yield Gains in Five Major Crop Plants*, Madison: Crop Science Society of America, pp. 89–101.

Smale, M. (1997), 'The Green Revolution and Wheat Genetic Diversity: Some Unfounded Assumptions', *World Development*, 25, 1257–69.

Southgate, D. (1997), 'Alternatives to the Regulatory Approach to Biodiverse Habitat Conservation', *Environment and Development Economics*, 2(1), 106–10.

Srivastava, J., N. Smith and D. Forno (1996), 'Biodiversity and Agriculture. Implications for Conservation and Development', World Bank Technical Paper no. 321; World Bank, Washington, DC.

Swanson, T. (ed) (1995), *The Economics and Ecology of Biodiversity Decline. The Forces Driving Global Change*, Cambridge: Cambridge University Press.

Swanson, T. and T. Goeschl (1999), 'Evolving Property Right Regimes Regarding Plant Genetic Resources', *International Journal of Biotechnology*, 1(3), 55–87.

Thirtle, C.G. (1985), 'Technological Change and Productivity Slowdown in Field Crops: United States, 1939–1978', *Southern Journal of Agricultural Economics*, 17, 33–42.

Thirtle, C.G., V.E. Ball, J.C. Bureau and R. Townsend (1994), 'Accounting for Efficiency Differences in European Agriculture: Cointegration, Multilateral Productivity Indices and R&D Spillovers', mimeo, University of Reading.

Traxler, G., J. Falck-Zapeda, J.I. Ortiz-Monasterio and K. Sayre (1995), 'Production Risk and the Evolution of Varietal Technology', *American Journal of Agricultural Economics*, 77, 1–7.

Tzotzos, G. (1995), 'Genetically Modified Organisms: A guide to Biosafety', Wallingford: CABI.

Virchow, D. (1999), *Conservation of Genetic Resources: Costs and Implications for a Sustainable Utilization of Plant Genetic Resources for Food and Agriculture*, Berlin: Springer Verlag.

White, F.C. and J. Havlicek (1979), 'Rates of Return to Agricultural Research and Extension in the Southern Region', *Southern Journal of Agricultural Economics*, 11, 107–11.

World Conservation Monitoring Centre and Faculty of Economics (1996), 'A Survey of the Plant Breeding Industry', in T. Swanson and R. Luxmoore (eds), *Industrial Reliance Upon Biodiversity*, Cambridge: WCMC.

8. Forecasting the impact of genetic use restriction technologies: a case study on the impact of hybrid crop varieties

Timo Goeschl and Timothy Swanson

INTRODUCTION

There is a general expectation among proponents of biotechnological modification of crops that advanced techniques of genetic manipulation in plants will significantly enhance agricultural productivity. Concurrently, this technological development has delivered the means of protecting the genetic information responsible for these productivity improvements against unauthorized reproduction and extraction. These technologies have been termed 'genetic use restriction technologies' (GURTs) in reference to the control of unauthorized use of the novel genetic structure that delivers improved traits.[1]

Although still at the patent stage, there are several areas of concern about the potential implications of genetic use restriction technologies. Some observers worry about the environmental effects of gene flow from crops thus sterilized to other plants, causing the potential sterilization of seeds beyond the confines of the individual field (Jefferson *et al.*, 1999; Crouch, 1998). The distribution of economic rents between farmers, seed companies and consumers is another area of possibly undesirable consequences (Srinivasan and Thirtle, this volume). Others are concerned about the impacts of these technologies on the livelihoods of subsistence farmers that predominantly rely on saved seed for replanting their fields. The focus of this chapter is the analysis of genetic use restriction technologies from the perspective of agricultural productivity growth through crop-improving innovations and the diffusion of these innovations to developing countries.

This focus situates the problem in the context of economic development and agricultural R&D. In the management of the research and development

(R&D) process, society is attempting to solve a particular form of a public goods supply problem. The information generated by the R&D process has the character of a public good; that is, it is non-rival and non-excludable. In the absence of regimes that ensure a flow of rents to the creator of this information, its public good character poses a strong disincentive for private investment in R&D. The rationale for the creation of property rights in innovations is that such regimes will have the effect of encouraging investments in R&D, and hence the supply of information resulting from it. Plant breeding forms an essential part of R&D in the agricultural sector. It has been shown to be a major source of agricultural productivity growth (Traxler *et al.*, 1995; Huffman and Evenson, 1993; Schmidt, 1984; Thirtle, 1985; Evenson and Kislev, 1973). At the same time, plant breeding poses an even more formidable problem to society than usual R&D processes because the R&D output, namely seeds, has a self-reproducing property. This makes it very difficult for the innovator to control the dissemination of innovative traits. Additionally, cross-breeding offers competitors the potential for accumulating others' innovations within their own R&D output. In essence, the ease of transfer of traits between crops makes it very hard to protect the proprietary information contained in improved varieties (Swanson, 1996). The result is that, in the absence of intellectual property rights, very little private R&D would be carried out in plant breeding in comparison to the social benefits that are generated through such crop improvements.

The fundamental idea is that GURTs represent a novel mechanism for appropriating the rents from innovation in the plant breeding industry. This mechanism radically enhances the plant breeder's scope for rent capture. This is likely to result in increased private investment in agricultural R&D and hence in a higher rate of innovation in the plant breeding industry. On the other hand, a consequence of this novel mechanism is that it significantly complicates the dissemination of crop improvements through adaptive breeding and informal seed trade.

Technologies that inhibit the dissemination of innovations are likely to have adverse impacts on countries for which innovations created abroad are the major source of crop improvement and that are therefore dependent on access to this flow for productivity increases.[2] Among these countries are the least developed of the world (Coe *et al.*, 1997). One of the possible consequences of GURTs is hence that they may lead to a distinct downward shift in the growth trajectories of agricultural productivity in developing countries. This downward shift would work in the opposite direction of the potential increase in private R&D for crop improvement stimulated by GURTs. It would be caused by restrictions in the flow of innovations to which developing countries have had access over the period of pronounced yield growth in the last 50 years. If this restriction outweighs the effect of increased R&D, developing countries are likely to face cumulative losses in agricultural productivity growth as a result of widespread adoption of GURTs by crop innovators.

How can the possibility of this link between GURTs and a reduction in the rate of diffusion be substantiated? In this chapter, we draw on the experiences with a previous use restriction technology in agriculture in order to establish that the mechanism for rent appropriation in agricultural R&D has measurable effects in terms of both the rate of innovation at the technological frontier and the rate of diffusion from the frontier to developing countries. This empirical evidence allows us to make simulation-based inferences regarding the probable impact of GURTs on developing countries and on the diversity of experiences that this technology will bring about among these countries, based on structural differences in their capacity to capture the flow of innovations.

Although GURTs are of recent origin, technologies with similar characteristics have existed for many decades, specifically the hybridization of cultivated varieties. This technique has been available for commercial seeds since the 1920s. Hybridization of cultivars has two implications: one is that the replanting of seeds from a hybrid results in a rapid deterioration of yield potential,[3] the other is that it protects against unauthorized reproduction by farmers and that the composition of the hybrid can be withheld from other breeders if the innovator does not disclose the inbred lines that make up the hybrid crop. GURT crops share these two characteristics with hybrid crops, albeit in more extreme forms: replanting of GURT seeds results in an expected yield loss of close to 100 per cent, and the reproduction of the crop's underlying genetic structure by a third party is currently not feasible since reproduction of the seed itself is impossible (DfID, 1999). The application of hybridization in the commercial seed sector also suggests that the availability of use restriction through this technique has been widely used by private companies when investing in R&D in those crops in which hybrids are feasible (Butler and Marion, 1985; Butler, 1996). This means that hybrid crops share fundamental features of use restriction with GURTs (although these features operate to different degrees of perfection in these two applications) and that industry has made significant use of these features as a form of rent protection.

In a previous publication (Goeschl and Swanson, 2000), we have estimated the rate of diffusion of innovations in hybrid and non-hybrid crops over the last 40 years. These estimates can be regarded as indicative of the likely impacts of the adoption of GURTs by crop innovators. Here we use these estimates as the basis for forecasting what probably constitutes the lower bound on the impact of GURTs in developing countries. These impacts are expected to arise out of the application of GURTs to those crops for which hybridization is currently not carried out on a significant scale, such as wheat and rice.

The remainder of the chapter is organized as follows. The following section reviews the results of a 39-year panel study of yield developments in the most important hybrid and non-hybrid crops. In the third section, we apply these results in order to forecast the yield development in currently non-hybridized crops in selected countries. The likely development in the absence of GURTs, that is a perpetuation of the current regime, is then compared with the expected growth of yields when GURTs are widespread.

REVIEW OF PANEL STUDY ON DIFFUSION

In this section, we survey the results of a panel study on yield development in the eight most widely cultivated crops,[4] barley, cotton, maize, millet, rice, sorghum, soybeans and wheat, covering 39 years, from 1961 to 1999. For a full description of the data, econometric methodology and modelling, we refer to Goeschl and Swanson (2000 and 2001). In the past, use restriction has been crop-specific: in two of the eight crops, namely maize and sorghum, the vast majority of improved varieties are hybridized. In the remaining six, hybridization is rare. This crop specificity enables us to compare the performance of hybrid and non-hybrid crops with respect to diffusion.

The method used is a fixed-effect panel estimation model that allows for heterogeneity among the countries through variable intercepts. The specific model that is being estimated is common in the empirical estimation of productivity convergence in other sectors and widely used in the literature on economic growth (Barro and Sala-i-Martin, 1995).[5] This literature considers the diffusion of technology as a continuous process of innovations occurring at the productivity frontier and subsequently diffusing to developing countries.[6] This process is modelled as a sequence of exogenous stochastic stocks, that is, the event of an innovation, that sets countries back in their relative yields in comparison to the technological frontier.[7] We test for the presence of a differential in the rate of diffusion through a dummy variable for observations involving a hybrid crop. The model then estimates for each category the rate at which this shock is compensated for, allowing for heterogeneity in the intrinsic 'rate of recovery' between countries. The model has the form

$$\Delta G_{it} = a_i + \beta \cdot G_{i,t-1} + \gamma \cdot D \cdot G_{i,t-1} + \epsilon, \tag{8.1}$$

where G is the gap (difference) in logarithm between the yields in a specific country and the lead country and Δ signifies the change in the gap. The intercept term a_i denotes the long-term difference in productivity growth in equilibrium. One way of interpreting a_i is to regard it as a country-specific intercept that captures the agroecological and institutional factors that influence the overall productivity development of the country. In this it captures the content of the hypotheses that claim that country-specific factors are responsible for the disproportionate yield gap that exists in the case of maize and sorghum. The coefficient β that is to be estimated then reports the diffusion coefficient across all crops and γ is the diffusion rate differential for hybrid crops identified through the dummy variable D.[8] Empirically, we perform Fisher's test as proposed by Maddala and Wu (1999) as a panel data unit root test. We then estimate the diffusion coefficient β and the diffusion rate differential γ according to equation (8.2) as a GLS-regression correcting for the residuals being cross-section heteroscedastic by down-weighting each pool equation by an estimate of the cross-section residual standard deviation.[9]

Econometric Results

The estimation delivers coefficients β and γ that are statistically highly significant. We also report the *average* intercept for all countries in the estimation denoted by $\hat{a}$. Before interpreting the results, it is convenient to perform some algebra in order to bring the model into a simpler form. Rearranging (8.1), we arrive at the following equation for the growth rate of yield, $\Delta\hat{y}_t$ in the average developing country:

$$\Delta\hat{y}_t = \Delta y_t{}^* - (1 + \beta + \gamma \cdot D) \cdot G_{t,t-1} + \hat{a} + \epsilon. \tag{8.2}$$

This formulation highlights the separate components that drive the growth rate of yields in the average developing country. The first component is the yield gain at the frontier Δy^*. This reflects the expansion of the set of technological possibilities. The second component captures the extent to which an innovation can diffuse in the country. We define the gap G to take on positive values. Therefore we would expect that the coefficient β is negative (indicating that innovations do not have a negative effect on growth) and that the closer the coefficient is to -1, the more rapidly the gains dissipate from the frontier to the average developing country. The third component, γD is the effect of hybridization on the growth rate. The fourth parameter, $\hat{a}$, summarizes the country-specific growth lags as an average. A positive value would indicate that, on average, developing countries have a higher 'intrinsic' rate of yield growth in this crop and vice versa.

Interpreting the Results: Diffusion

Table 8.1 shows the results of the econometric estimation of equation (8.1). The most important result is that hybridization has a measurable impact on the rate of diffusion. The coefficient of the hybrid dummy variable is highly significant, despite allowing for fixed effects both by country and by crop. The rate of diffusion of innovations from the frontier to developing countries across all crops was that such crops carried over roughly 69 per cent of the gap opened by an innovation into the next year. The 'diffusion penalty' involved in having innovations predominantly occur in hybridized crops is about 7.1 per cent per year. This means that developing countries retained about 7 per cent more of the yield gap each year in hybrids than in non-hybrids. This explains an important part of the cumulative yield gap that has developed in hybrids. The results also indicate that there is merit to the idea that structural effects, such as agroecological conditions, have contributed to inhibiting yield growth of hybrids in developing countries. The parameter $\hat{a}$ is the mean of the individually estimated parameters a_i. The means computed for hybrids and non-hybrids indicate that, in hybrids, the average developing country has had a greater negative long-term deviation from the growth rate of the frontier than in non-hybrids. The combination of structural and diffusion effects is therefore responsible for the significant gap in yields that persists between developed and developing countries in hybrid crops.

Table 8.1 Regressions for diffusion of innovations in different crops

Coefficient	Means and standard variation
β (diffusion – all crops)	−0.313
	(0.008)***
γ (diffusion – hybrids)	0.071
	(0.011)***
α	−0.33611
R^2	0.16
(number of observations)	(14 858)
DW-statistic	2.39

Note: The figures in parentheses is the standard error; *** at 1% level.

The results on the rate of diffusion have an intuitive economic interpretation: for a farmer cultivating different crops, an important criterion for evaluating crops is the loss of yield suffered as a result of slow diffusion. This loss can be assessed as the present value of the cumulative process of an innovation arriving at developing countries in a delayed fashion rather than arriving immediately.[10] Figure 8.1 reports the multiplier to the initial shock in order to estimate the present value of the loss at a 10 per cent discount

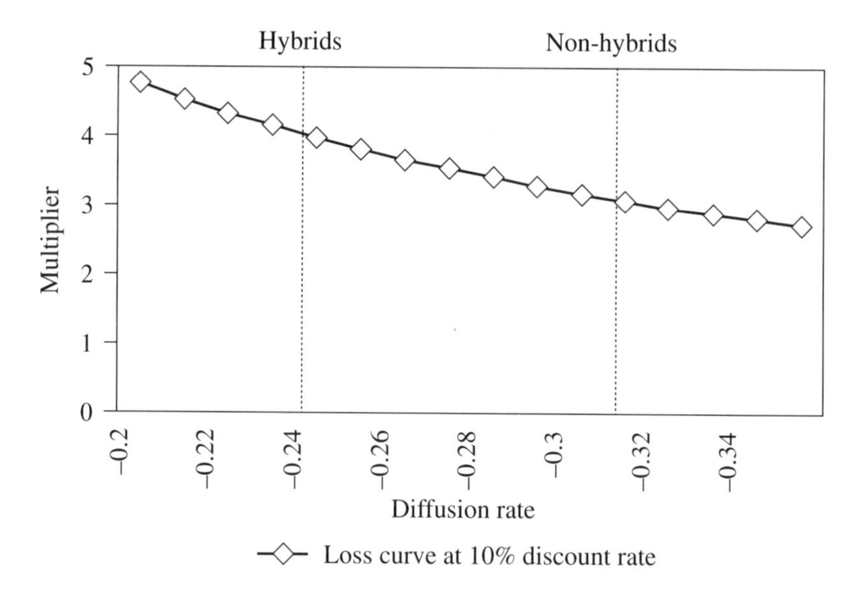

Figure 8.1 Loss multiplier as a function of the diffusion rate

rate.[11] The interpretation of Figure 8.1 is the following. Consider an innovation at the frontier in period 1 that results in an increase in profits by, say, 100 dollars for the average farmer. Within this period, the farmer in the developing country does not receive any of the innovation, thus incurring a loss of 100 dollars. In the next period, the first benefits start to trickle down from the frontier, thus decreasing the loss relative to the frontier, with more of the benefits becoming available in subsequent periods. The curve depicts the present value of the total accumulated losses as a factor that can be applied to the initial 'loss'. Figure 8.1 shows the economic loss of the frontier shifting away from the developing country by virtue of an innovation at the frontier as a function of the rate of diffusion. Based on the parameters estimated for hybrids and non-hybrids, the discounted cumulative loss in the case of hybrids is about four times the initial shock, while for non-hybrids it is around three times. This implies cumulative losses being about a third higher in hybrids than in non-hybrids.

Interpreting the Results: Country-specific Lags

A second set of important differences arises from the country-specific data on 'individual growth capacity'. The first result is that, on average, developing countries would experience slower growth in all crop yields as the coefficient $\hat{a}$ is below zero for all crops. However, these impediments to growth are quite different between crops, ranging from rice, a crop with good intrinsic growth potential in developing countries at $\hat{a} = -0.230$, to wheat, with high average barriers to growth at $\hat{a} = -0.384$. This captures whether innovations have been diffusing to countries where the local conditions are beneficial or adverse to the successful cultivation of the plant.[12] Interestingly, there is no correlation between parameter estimates of $\hat{a}$ and β, which indicates that the processes of diffusion are disjoint from the effects of local conditions.[13]

FORECASTING THE IMPACT OF GENETIC USE RESTRICTION TECHNOLOGIES

With respect to its impact on the unauthorized reproduction of advanced cultivars, hybridization is a technological precursor of genetic use restriction technologies. It shares fundamental features with GURTs, although these features are exhibited in GURTs to a higher degree of 'perfection' than in hybrids.

Crops that have so far not been marketed as hybrids are the most likely target of genetic use restriction technologies (DfID, 1999). It will be in these crops, therefore, that the impact of GURTs will be the most significant as the regime of intellectual property protection will shift from an essentially public domain provision to a use-restricted regime. On the basis of the empirical estimates on hybrids, we carry out some forecasts regarding the likely impact of adopting use restriction technologies in these crops.

Baseline Scenario and Parameters

The panel study report is based on a leader–follower framework of innovation and diffusion (Barro and Sala-i-Martin, 1995). The simulation extends this framework into a forecasting situation where the three components determining the growth of yields in the developing country are the expansion at the frontier, the rate of diffusion from the frontier to the developing country and the long-term differences in growth rates in crop yields. The baseline scenario for this forecast is the continued absence of GURTs from seed production. This baseline is established by assuming a perpetuation of the estimated growth rate at the frontier and rate of diffusion from the frontier. We refer to the baseline as the scenario 'in the absence of use restriction'.

The experience from hybrid crops suggests that expansion at the frontier will benefit from increased scope for rent capture by stimulating private

R&D (Srinivasan and Thirtle, this volume). We therefore assume that the rate of expansion at the frontier will be higher in the presence of GURTs. This is important because it implies that *ultimately* every developing country's yields will be higher under use restriction than under the continuation of the present public domain regime. This underlines the importance of GURTs providing private industry incentives for the long-term development of yields. Owing to the presence of a discount rate, however, the welfare effects of GURTs are not time-invariant. The crucial question is *how long* it takes for yields under GURTs to overtake the baseline scenario without use restriction. This is determined by the rate of diffusion, the country-specific long-term deviation from the growth rate at the frontier and the current gap in yields between the individual developing country and the frontier.

Our forecasting simulation looks at a 20-year time horizon for the development of yields in non-hybrid crops under the two scenarios. In order to simplify the comparison, the average values of the six non-hybrid crops are used.

For the productivity frontier, we assume a starting yield in non-hybrids of 4t/ha. for the year 2000 in both the baseline and the use restriction scenario.[14] For the baseline scenario, we assume a continued growth at 1.58 per cent per annum in the yields of non-hybrid crops in developed countries. This is the mean of past growth rates in the six different crops at the frontier (see Goeschl and Swanson, 2001). For the use restriction scenario, we initially assume that yield growth will take place at the rate experienced in hybrids in developed countries over the last 40 years, which has been 2.175 per cent on average (ibid.). The parameters estimated through equation (8.1) and reported in Table 8.1 are used as parameters for the rates of diffusion under the baseline (−0.312) and the use restriction (−0.242) scenarios. Adoption of GURT crops in developing countries is assumed not to take place as long as there are no yield benefits from the technology.

Simulation results: all developing countries

Figures 8.2 and 8.3 show the yield histogram and yield statistics for all 86 developing countries analysed in this chapter under the baseline scenario and that of widespread use of GURTs in the year 2020. These figures summarize the overall impact of GURTs relative to a continuation of the current regime.

In the absence of changes to the current IPR regime, the average yield in developing countries will be 23.1 tons per hectare at that point in time. If GURTs were adopted, average yields would be about 8 per cent (or about 2 tons) lower. By contrast, average yields in developed countries after widespread adoption of GURTs are forecast to exceed the baseline by more

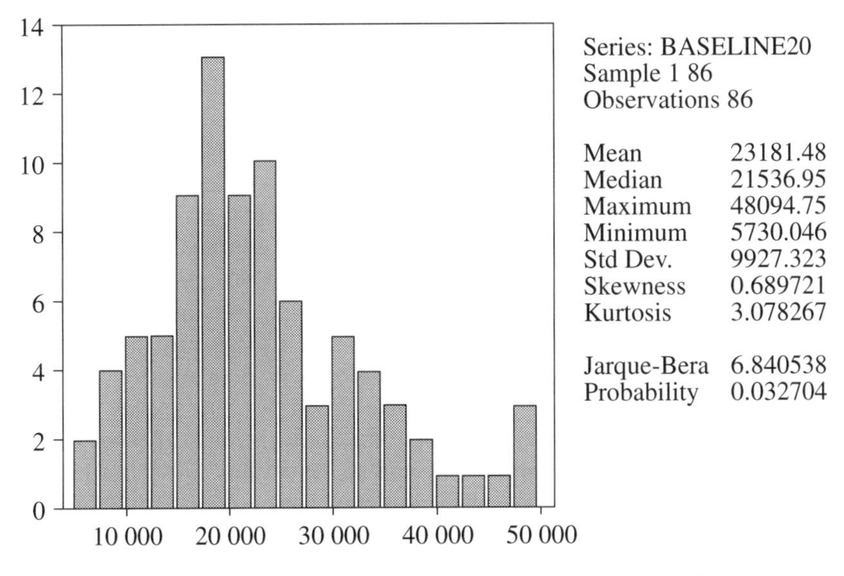

Figure 8.2 *Yield histogram and yield statistics for the year 2020 for 86 developing countries in the baseline scenario*

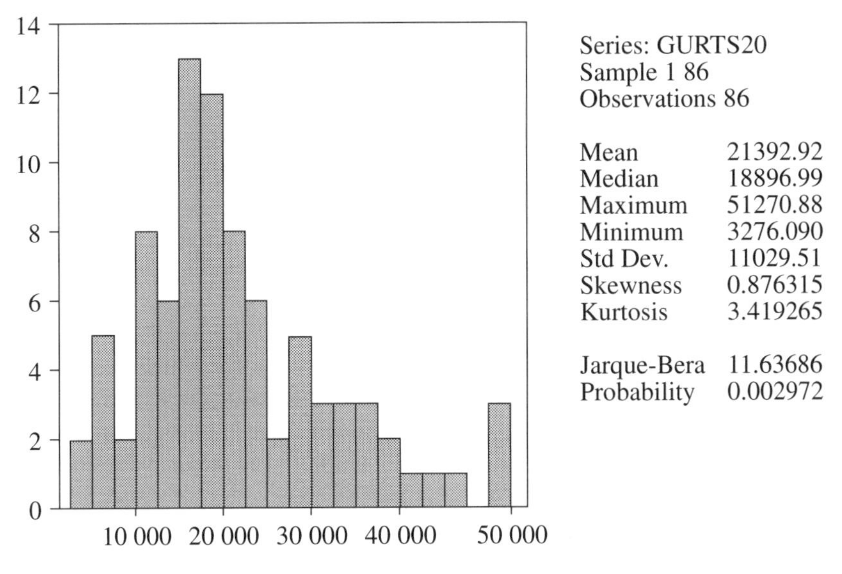

Figure 8.3 *Yield histogram and yield statistics for the year 2020 for 86 developing countries in the GURTs scenario*

than 10 per cent by 2020. It is only after another 20 years, that is in the year 2040, that developing countries' average yields under GURTs overtake those under the baseline regime.[15]

Another aspect is the distributive effects of GURTs between developing countries. One indicator of this is the coefficient of variation in yields across developing countries. In the baseline scenario, this coefficient is 0.42; in the case of GURTs it rises to 0.52, indicating that under GURTs differences in agricultural productivity will increase rather than decrease. At the general level, then, the two principal conclusions are therefore that GURTs tend to lead to an initially flatter growth curve and hence lower yields in developing countries over a 20-year period *on average* and that the variance in yields, that is, distributive disparity, will increase.

The increasing variance in developing country yields merits further examination since it suggests that individual countries will experience very different results of an adoption of GURTs. To do this we selected three developing countries out of the 86 countries sampled in order to illustrate some of the diversity of outcomes. These countries are China, Ethiopia and Tanzania (Table 8.2). They were selected on the basis that all crops included in the sample are grown there, thus providing better data, and that they represent the widely divergent experiences among developing countries.

Table 8.2 Country-specific simulation parameters

	China	Ethiopia	Tanzania
Average yield in non-hybrids as percentage of developed country yield	85.1	56	25
Average country-specific long-term deviation from developed countries' growth rate in non-hybrids	−0.094	−0.2128	−0.338

Table 8.2 reports the country-specific parameters that enter into the simulations. Yields in the initial period are below developed countries' yields by the average yield gap in non-hybrids. Across these six crops, China has the lowest average shortfall in yields relative to developed countries with a 15 per cent gap, while Ethiopia has a little over half the yields of developed countries. Tanzania does particularly poorly with a gap of about 75 per cent. What is also important for the simulation is the country's long-term deviation from the yield growth rate in developing countries across the six crops. These data are generated by the estimation of equation (8.2) on a crop and country-specific basis.

Simulation results: selected countries

Figures 8.4 to 8.7 report the simulation output graphically. The forecasts show that the individual country experiences vary quite considerably. In developed countries (Figure 8.4), the adoption of user restriction results in higher growth rates in yield and a more favourable yield development over the 20-year time horizon. There are developing countries where the experience is similar to that of developed countries, but arises in a more delayed fashion. In China (Figure 8.5) for instance, yields in the first ten years are expected to be very similar under both scenarios before the impact of use restriction on the yield frontier begins to push yields in China above the baseline. The case of Ethiopia (Figure 8.6) illustrates a country that in the short run would be better off under the current regime, as the flow of innovations would be more easily appropriable. However, towards the end of the 20-year horizon, the faster expansion at the technological frontier under user restriction has compensated for the slower diffusion inherent in this regime. Lastly, the case of Tanzania (Figure 8.7) illustrates a case where, for the foreseeable future, the country would be worse off under a use restriction scenario than under a perpetuation of the current regime.

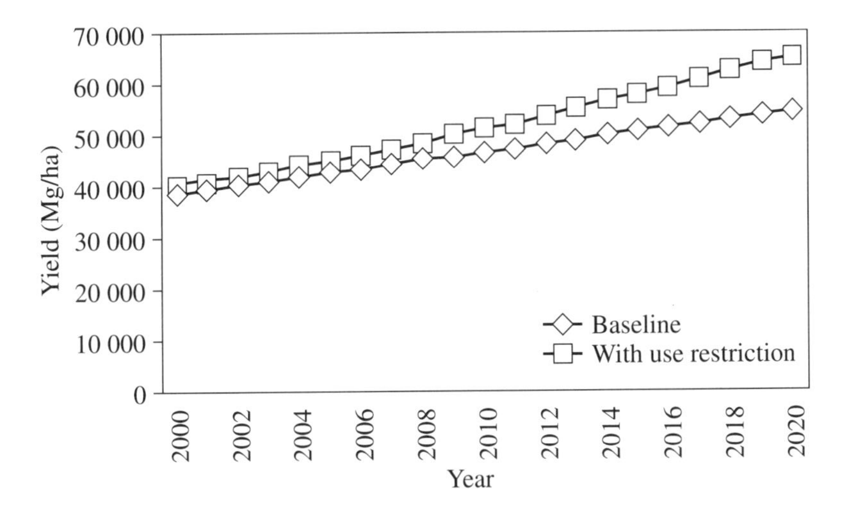

Figure 8.4 *Comparison of yields under the use restriction and baseline scenarios, developed countries, 2000–2020*

These four cases illustrate the diversity of outcomes that can be expected as a result of a potential adoption of genetic use restriction technologies. This diversity implies that, over a policy-relevant time horizon of 20 years, countries will not be indifferent as to the regime adopted, depending on the current state of a country's agriculture. The simulations suggest that

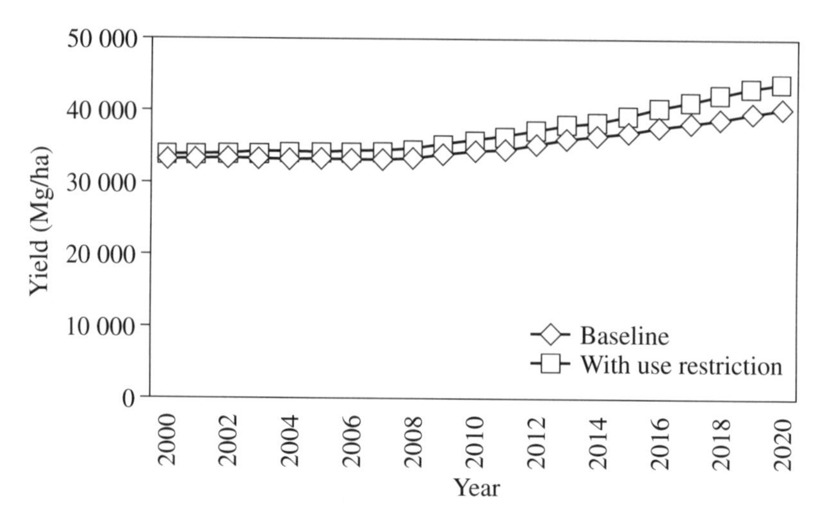

Figure 8.5 Comparison of yields under the use restriction and baseline scenarios, China, 2000–2020

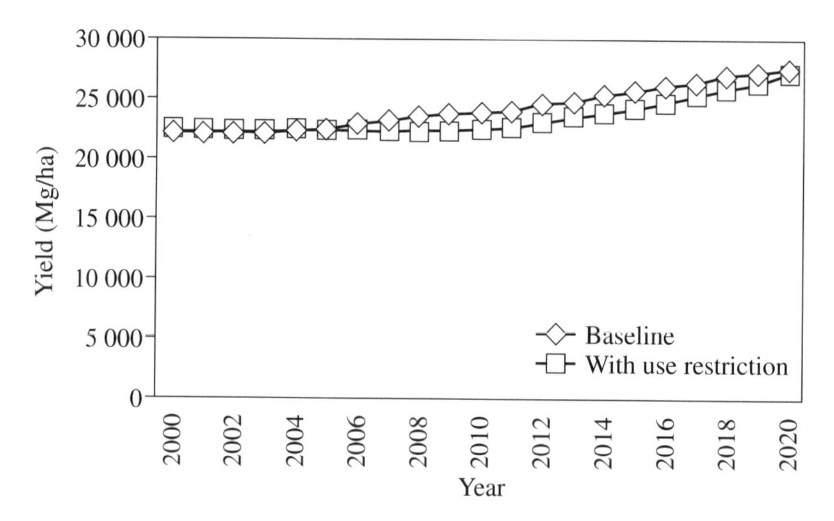

Figure 8.6 Comparison of yields under the use restriction and baseline scenarios, Ethiopia, 2000–2020

the most advanced countries stand to benefit most from use restriction while the least advanced stand to lose. As stressed before, when projected sufficiently far into the future, the productivity gains that the stimulation of private R&D through use restriction delivers result in the baseline scenario being overtaken. However, the net present value of these gains may

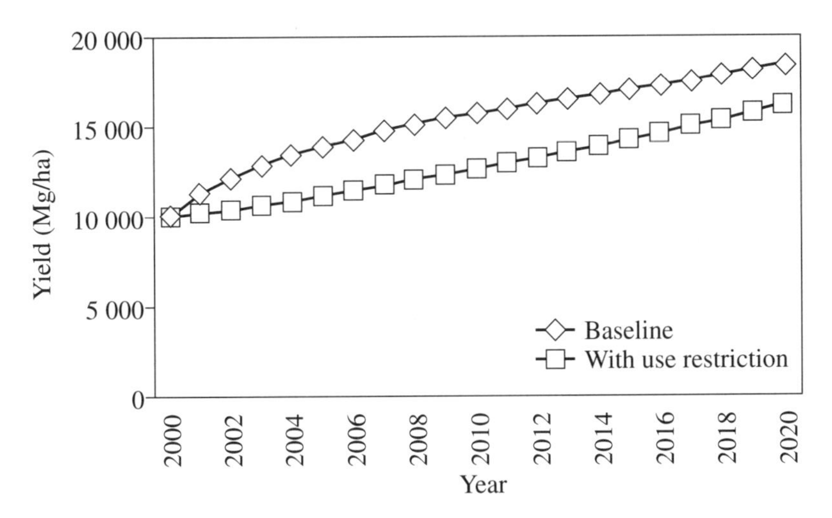

Figure 8.7 *Comparison of yields under the use restriction and baseline scenarios, Tanzania, 2000–2020*

currently be insufficient for developing countries to outweigh the short-term losses. It is interesting to note that, even if GURTs led to a doubling of the rate of innovation seen in hybrids at the same rate of diffusion, it would take more than ten years in the case of Tanzania for yields under use restriction to outperform the baseline yields.

Some Implications

The shift in the growth trajectory that developing countries are likely to experience as a result of a widespread adoption of GURTs will lead in the long run to higher yields everywhere. However, most countries, and particularly the least developed ones, will have to pass through a phase of loss relative to the present regime for the diffusion of agricultural innovations. One of the implications is that this will lead, at least in the short to medium term, to the emergence of new 'growth clubs' in agricultural development as countries will be put on different yield growth trajectories. The distributional consequences of this development over time may be considered undesirable. Another implication is that developing countries will need to develop new approaches to capturing the results of international agricultural R&D in order to maintain the flow of innovations into the country and thus to mitigate the adverse transitional period to higher yield growth. This leads to the wider question about the role of the public sector at a national and international level when use restriction technology is widespread. For some countries, such as in the case of Tanzania, new technological, contractual and regulatory measures would be required in order to avert significant welfare losses.

Qualifications

There are a number of qualifications to this forecast, some of which are generic to any forecasting exercise, some of which are case-specific. The generic qualifications are that any ex ante appraisal of a technology has a wide error margin as there is no observable evidence from the technology per se and that data available on antecedent technologies have their limitations. Also the simulation describes a world where the public response to the widespread adoption of GURTs mirrors the historical response in the case of hybrids. This need not be the case and this simulation exercise is intended to invite policy makers to reflect on the effects of previous policy responses. Specific qualifications are that the country impacts estimated are unweighted averages across all non-hybrids. This tends to play down the impact of crops where a country has a comparative advantage in production. Also the simulation does not take into account any endogenous adjustments that might take place as a result of use restriction technology becoming available. Obvious examples would be a change in the portfolio of crops produced. It is not clear, however, whether including such adjustments would necessarily decrease the error margin. The final qualification is that the simulation is based on empirical estimates derived from a comparative study of innovation and diffusion in hybrids and non-hybrids. Although the *direction* of the impact of GURTs is likely to be the same as the impact of hybrids, the *volume* of the impact is harder to assess. Since GURTs present a more advanced form of use restriction, the impact of hybrids is likely to represent the lower bound on the impact of GURTs. Interestingly, though, the timing of the impacts illustrated by Figures 8.2 to 8.5 is not significantly altered if there is a parallel decrease in the rate of diffusion when the rate of innovation increases.

CONCLUSION

If, as it is reasonable to expect, GURTs replicate the experiences in hybrid-based agriculture across all other staple crops, this will lead in the first instance to a higher rate of investment by private industry in crop improvement motivated by enhanced scope for rent capture. This could result in a net increase or decrease in the amount of crop improvement produced globally, depending on the degree to which this private spending will crowd out public expenditures on agricultural R&D. We have assumed for the purpose of this simulation that the rate of innovation will rise to the level experienced in hybrid crops over the last 40 years. The impact of GURTs on the rate of diffusion of these new innovations is less ambiguous. On the basis of the experiences with hybrid crops, we can predict that GURTs will have a negative impact on the rate of diffusion of innovations in those crops for which developing countries could previously rely on an inflow of innovations from abroad. This is because the advanced germplasm incorporated within advanced commercial cultivars will be much more costly to extract, reproduce and disseminate, impeding the adaptation of the latest generation of value-adding traits to local farming systems. This means that developing countries that predominantly use varieties that are currently not hybridized will experience a discontinuous shift onto a different trajectory of yield growth from the one they are currently moving along. For most countries, this implies a period of deteriorating net growth relative to the current regime. It is distributionally problematic that this period is the longest for the least developed countries. It is for this reason that GURTs present a challenge to the global regulation of biotechnologies and to the role of the public sector in generating and diffusing productivity gains.

ACKNOWLEDGMENTS

This research has its origins in a research grant by the Department for International Development (CNTR 99 8215) to investigate the impacts of genetic use restriction technologies on developing countries. We are grateful to Robert Carlisle from DfID for encouraging us to explore this area. We are particularly grateful to James Symons for helpful discussions on the econometrics and comments, and grateful to Mark Rogers for helpful discussions and comments, without implicating them in any way in all the remaining errors. This research has benefited from discussions with Ed Barbier, William Fisher, Jonathan Jones, Michael Lipton, C.S. Srinivasan and Colin Thirtle that took place as part of the research.

NOTES

1. In the general public, they have become better known by the name 'terminator genes'. This epithet was coined by Rural Advancement Foundation International, an NGO; 'traitor technology' has been another suggestion.
2. The national agricultural research centres (NARCs) and the Consultative Group on International Agricultural Research (CGIAR) system currently provide mechanisms for facilitating this flow of innovations through adaptation of advanced material to local conditions.

3. The first-generation loss is normally in the order of 25 to 30 per cent.
4. The criterion applied here is the global acreage of a crop.
5. For a full development of the model in the context of fixed effects such as agroecological factors, see Goeschl and Swanson (2000).
6. This type of analysis has to be contrasted with earlier studies of technological change such as Griliches (1957) where a discrete innovation is examined as it diffuses intertemporally and spatially from its origin.
7. The yield data for the eight crops examined are annual yield data from the FAO Statistical Database (FAOSTAT). The data record the harvested production per unit of harvested area for crop products based on the annual harvest data and the area harvested. Data are recorded in hectogrammes (100 grammes) per hectare (HG/HA). The data are not always fully reliable. Specifically, all countries were omitted for which no complete time series of yield data was available for the 39-year period or whose yield data showed an obvious lack of reliability. Even with a stringent application of these rules, the panel size is never below 39 countries, with soybeans particularly affected by the rule of complete time series. The frontier is defined by the average yield in developed countries.
8. For observations involving hybrid crops, $D=1$.
9. The presence of heteroscedasticity tends to lead to higher diffusion coefficients. This weighting procedure corrects for that. The White test for cross-section heteroscedasticity is performed for all estimations and reports consistent parameters for all crops.
10. This curve is constructed on the assumption that the demand curve for agricultural output has constant and equal demand elasticity in the developed and developing countries.
11. The curve is fairly robust against changes in the discount rate. A higher rate pushes the curve down slightly as future losses have a lower value, and vice versa.
12. There are for each crop countries in which the intrinsic growth rate of the yield is basically equal or above that prevalent in the frontier countries. In the case of barley, this holds for Zimbabwe; in the case of cotton, for Israel and Syria; in the case of maize, for Chile; in the case of millet, for China; in the case of rice, for Egypt and Korea; in the case of sorghum, for Egypt and Israel; in the case of soybeans, for Ethiopia; and in the case of wheat, for Egypt and Zimbabwe.
13. The correlation coefficient between $\hat{a}$ and β is 0.02 across all crops.
14. The simulation is not sensitive to particular numerical yield values. The initial yield is only chosen for illustrative purposes.
15. This should be taken as indicative only, as the error of margin in the forecast increases dramatically from 20 to 40 years.

REFERENCES

Alston J.M. and R.J. Venner (1998), 'The Effects of U.S. Plant Variety Protection Act on Wheat Genetic Improvement', paper presented at the symposium on 'Intellectual Property Rights and Agricultural Research Impact' sponsored by NC 208 and CIMMYT Economics Program, 5–7 March, El Batan, Mexico.

Alston J.M., P.G. Pardey and V.H. Smith (1998), 'Financing Agricultural R&D in Rich Countries: What's Happening and Why?', *The Australian Journal of Agricultural and Resource Economics*, 42(1), 51–82.

Barro, R.J. and X. Sala-i-Martin (1995), *Economic Growth*, New York: McGraw-Hill.

Butler, L.J. (1996), 'Plant Breeders' Rights in the U.S.: Update of a 1983 Study', in J. Van Wijk and Walter Jaffe (eds), *Proceedings of a seminar on 'The Impact of Plant Breeders' Rights in Developing Countries', held at Santa Fe Bogota, Colombia*, 7–8 March 1995, Amsterdam: University of Amsterdam, pp. 17–33.

Butler, L.J. and B.W. Marion (1985), *The Impact of Patent Protection on the U.S. Seed Industry and Public Plant Breeding*, Food Systems Research Group Monograph 16, University of Wisconsin Madison, Madison, WI.

Capalbo, S.M. and J.M. Antle (1989), 'Incorporating Social Costs in the Returns to Agricultural Research', *American Journal of Agricultural Economics*, 71, 458–63.

CIMMYT (1999), *A Sampling of Impacts 1999. New Global and Regional Studies*, Mexico: CIMMYT.

Coe, D.T., E. Helpman and A. Hoffmaister (1997), 'North–South R&D Spillovers', *Economic Journal*, 107, 134–49.

Crouch, M. (1998), 'How Terminator Terminates', rev. edn, Edmonds Institute Occasional Papers, Edmonds Institute, Washington.

Dalton, T.J. and R.J. Guei (1999), 'Ecological Diversity and Rice Varietal Improvement in West Africa', a report submitted to the CGIAR impact assessment and evaluation group.

Department for International Development (DfID) (1999), *Costs and Benefits to the Livelihoods of the Rural and Urban Poor arising from the Application of so-called 'Terminator Genes' and Similar Technologies in Developing Countries*, report submitted by GS Consulting, London: DfID.

Evenson, R.E. (1989), 'Spillover Benefits of Agricultural Research: Evidence from US Experience', *American Journal of Agricultural Economics*, 71, 447–52.

Evenson, R.E. and Y. Kislev (1973), 'Research and Productivity in Wheat and Maize', *Journal of Political Economy*, 81(6), 1309–29.

Falck-Zepeda, J.B. and G. Traxler (1998), 'Rent Creation and Distribution from Transgenic Cotton in the U.S.', paper prepared for the symposium 'Intellectual Property Rights and Agricultural Research Impacts', sponsored by NC-208 and CIMMYT Economics Program, 5–7 March, CIMMYT Headquarters, El Batan, Mexico.

Frey, K.J. (1996), 'National Plant Breeding Study – I: Human and Financial Resource Devoted to Plant Breeding Research and Development in the United States in 1994', Special Report 98, Iowa Agriculture and Home Economics Experiment Station, Iowa State University, Iowa.

Fuglie, K., N. Ballenger and K. Day (1996), 'Agricultural Research and Development: Public and Private Investments Under Alternative Markets and Institutions', AER-735, Economic Research Service, United States Department of Agriculture.

Goeschl T. and T. Swanson (2000), 'The Diffusion of Innovations to Developing Countries: The Case of Crop Varieties', mimeo, Department of Land Economy, University of Cambridge.

Goeschl, T. and T. Swanson (2001), 'Genetic use restriction technology and the diffusion of yield gains to developing countries', *Journal of International Development*, 17(3), 58–91.

Griliches, Z. (1957), 'Hybrid Corn: An Exploration in the Economics of Technological Change', *Econometrica*, 48, 501–22.

Huffman, W.E. and R.E. Evenson (1993), *Science for Agriculture. A Long-Term Perspective*, Ames, IA: Iowa State University Press.

Jaffe, W. and J. Van Wijk (1995), 'The Impact of Plant Breeders' Rights in Developing Countries: Debate and Experience in Argentina, Chile, Colombia, Mexico and Uruguay', Inter-American Institute for Co-operation in Agriculture and University of Amsterdam, Amsterdam.

Jefferson RA. with D. Byth, C. Correa, G. Otero and C. Qualset (1999), 'Genetic

Use Restriction Technologies. Technical Assessment of the Set of New Technologies which Sterilize or Reduce the Agronomic Value of Second Generation Seed as Exemplified by U.S. Patent no. 5,723,765 and WO 94/03619', expert paper, prepared for the Secretariat of the Convention for Biological Diversity, Subsidiary Body on Scientific, Technical and Technological Advice.

Kalton, R.R. and P.A. Richardson (1983), 'Private Sector Plant Breeding Programmes: A Major Thrust in U.S. Agriculture', *Diversity*, 1(4), 16–18.

Kalton, R.R., P.A. Richardson and N.M. Frey (1989), 'Inputs in Private Sector Plant Breeding and Biotechnology Research Programs in the United States', *Diversity*, 5(4), 22–5.

Komen, J. and G.J. Persley (1993), 'Agricultural Biotechnology in Developing Countries: A Cross-Country Review', ISNAR *Research Report no. 2*, International Service for National Agricultural Research, The Hague.

Lesser, W. (1990), 'Sector Issue II: Seeds and Plants', in W.E. Siebeck (ed.), in *Strengthening Intellectual Property Rights in Developing Countries. A Survey of Literature*, Madison, WI: University of Wisconsin Press, pp. 59–68.

Maddala, G.S. and S. Wu (1999), 'A Comparative Study of Unit Root Tests with Panel Data and a New Simple Test', *Oxford Bulletin of Economics and Statistics*, Special Issue, 631–52.

Perrin, R.K., K.A. Kunnings and L.A. Ihnen (1983), 'Some Effects of the U.S. Plant Variety Protection Act of 1970', Economics Research Report no. 46, Department of Economics and Business, North Carolina State University.

Pray, C.E., M.K. Knudson and L. Masse (1993), 'Impact of Changing Intellectual Property Rights on US Plant Breeding R&D', mimeo, Department of Agricultural Economics, Rutgers University, New Brunswick.

Pray, C.E., S. Ribeiro and R.A.E. Mueller (1991), 'Private Research and Public Benefit – The Private Seed Industry for Sorghum and Pearl-Millet in India', *Research Policy*, 20(4), 315–24.

Schmidt, J.W. (1984), 'Genetic Contributions to Yield Gain in Wheat', in W.R. Fehr (ed.), *Genetic Contributions to Yield Gains in Five Major Crop Plants*, Madison: Crop Science Society of America, pp. 89–101.

Swanson, T. (1996), 'The Reliance of Northern Economics on Southern Biodiversity: Biodiversity as Information', *Ecological Economics*, 1–6.

Thirtle, C.G. (1985), 'Technological Change and Productivity Slowdown in Field Crops: United States, 1939–1978', *Southern Journal of Agricultural Economics*, 17, 33–42.

Thirtle, C.G., V.E. Ball, J.C. Bureau and R. Townsend (1994), 'Accounting for Efficiency Differences in European Agriculture: Cointegration, Multilateral Productivity Indices and R&D Spillovers', mimeo, University of Reading.

Traxler, G., J. Falck-Zapeda, J.I. Ortiz-Monasterio and K. Sayre (1995), 'Production Risk and the Evolution of Varietal Technology', *American Journal of Agricultural Economics*, 77, 1–7.

PART III

Biotechnology and Biodiversity: the Impacts of Biotechnologies on Conservation of Genetic Resources

9. Key issues in using molecular techniques to improve conservation and use of plant genetic resources

Carmen de Vincente, Toby Hodgkin and Geoffrey Hawtin

INTRODUCTION

Effective strategies to eliminate poverty, enhance food security and protect the environment require the continued maintenance and use of the genetic diversity present in the world's useful plant species. In its current strategy, the International Plant Genetic Resources Institute (IPGRI) emphasizes that conservation of plant genetic resources must strengthen their improved utilization. It recognizes that work on collecting and maintaining plant genetic resources needs to go hand-in-hand with work on the exploitation of diversity through germplasm enhancement and with improved deployment of traditional or enhanced germplasm to enrich production systems and increase their sustainability.

The dramatic advances in biotechnology over the past two or three decades have already had a substantial impact on many aspects of the conservation and use of plant genetic resources. Tissue culture technologies are important for the generation of new varieties and the provision and multiplication of disease-free plant materials for crop production (Dodds and Roberts, 1985; Barker and Torrance, 1997). Together with slow growth storage and cryopreservation (Engelmann, 1997), they are used routinely in many genebanks and play an important part in the safe movement of germplasm around the world (Barker and Torrance, 1997). Even more dramatic are the developments in molecular genetics that are having an increasingly profound impact on life sciences and agriculture and are changing fundamentally many aspects of plant genetic resources work.

THE POWER OF MOLECULAR GENETICS

Developing and Using the New Technologies

The past 15 years have witnessed rapid developments in molecular genetics. In 1984, the human immunodeficiency virus (HIV) became the first genome to be sequenced. In 1990, the international human genome-sequencing initiative was launched with an estimated duration of 15 years. By 1999, the Human Genome Project announced its aim to produce a 90 per cent complete sequence for spring 2000 and a 99.9 per cent complete version by 2003. By December 1999, the first human chromosome, Chromosome 22, had been fully sequenced (Dunham *et al.*, 1999). Such progress has been made possible principally by improvements in the equipment available and by the knowledge gathered through sequencing of organisms such as *Escherichia coli*, *Saccharomyces cerevisiae*, and *C. elegans*, the mouse genetic map and the *Drosophila* physical map as well. In plants, chromosomes 2 and 4 of *Arabidopsis thaliana* have already been completely sequenced (Lin *et al.*, 1999; Mayer *et al.*, 1999) and a first draft of the complete sequence of the rice genome has already been finished (Butler and Pockley, 2000).

At the same time, knowledge of the functions of DNA sequences has increased dramatically and improvements in technology, such as automation and robotics, have allowed the development of efficient methods for gene identification allowing broadening of the focus from the study of genetic variation to the functional analysis of the genome and as such to genome-wide activity. One of the most publicized developments has been that of 'DNA chips', micro-scale arrays of hundreds or thousands of DNA sequences that can be used to track patterns of gene expression in whole cells or organisms or to screen dozens to hundreds of samples for sequence variants.

The most significant application of molecular genetics in agriculture has been the development of genetically modified crops. Almost all crop species can now be transformed and, by 1998, well over 28 million hectares of genetically modified crops were being grown throughout the world (US Department of State, 1999, *http://www.csa.com/hottopics/gmfood/websites.html*). Nearly all transgenics released to date contain herbicide-resistant genes or the *Bacillus thuringiensis* (*Bt*) genes associated with insect resistance (*http://www.monsanto.com/monsanto/mediacenter/background*). The varieties have been largely produced for developed country agriculture, although significant adoption of transgenics has also occurred in China. Many countries view the use of transgenic crops with concern and there is currently a very vigorous debate on their use in agriculture, with a substantial number of countries refusing to permit their use in production fields.

Operating at the cutting edge in genetic technology entails large costs, and such research is increasingly done by the private sector. Genome sequencing has become a race between privately and publicly funded groups, with the added issue of whether the resulting sequences will be freely available or patent-protected. Multinational corporations dominate agricultural research and breeding, as smaller companies have been bought by a handful of larger ones. This has resulted in the consolidation of traditional breeding, seed supply, associated inputs such as fertilizers and herbicides, genomic research and transgenics under unified management. As a consequence there is growing concern over how such conglomerates will develop and distribute genetically modified food organisms, and the subsequent effects on biodiversity and social equity.

EMERGING OPPORTUNITIES FOR CONSERVATION AND USE

The new techniques and knowledge from molecular genetic studies are having a profound impact on the conservation and use of plant genetic resources. Molecular markers such as restriction fragment length polymorphisms (RFLPs), random amplified polymorphic DNA (RAPD), microsatellites (a simple sequence repeat, SSR) and amplified fragment length polymorphisms (AFLPs) detect variation in DNA sequence. Analysis of differences between plants as revealed by these markers is providing new insights in taxonomy and evolution. Measurements of marker variability can be used to identify populations with unique genetic properties or with high levels of genetic diversity. Combined with temporal and spatial analyses and socioeconomic studies, these approaches can provide a new understanding of the effects of environmental and human selection on the distribution of diversity at the DNA level.

Using molecular markers, genetic maps of great detail and accuracy have been developed. These have been used to identify quantitative trait loci (QTLs) for traits such as grain yield in maize and rice (Stuber *et al.*, 1992; Ribaut *et al.*, 1997; Austin and Lee, 1998; Xiao *et al.*, 1998, salt tolerance in tomato (Foolad and Jones, 1993; Foolad and Chen, 1999), and disease resistance in barley (Qi *et al.*, 1998, Pecchioni *et al.*, 1999), and to identify useful new genes controlling yield, crop quality or stress tolerance in crop wild relatives (Lindhout *et al.*, 1994; Eshed and Zamir, 1995; Bailey *et al.*, 1997; Monforte *et al.*, 1997; Foolad *et al.*, 1998; Grandillo *et al.*, 1999). Mapping studies have also revealed the inherent similarities ('synteny') between genomes (Ahn *et al.*, 1993; Ogihara *et al.*, 1994; Paterson *et al.*, 1995; Devos and Gale, 1997). The synteny between genomes allows specific useful genes identified in a well-studied species, including *Arabidopsis* (Liu *et al.*, 1996, Devos *et al.*, 1999), to be localized to the corresponding genomic site in other species that have not been studied in the same detail.

Markers such as RFLPs or AFLPs are 'anonymous' in the sense that they are not necessarily linked to any expressed trait. In contrast, expressed sequence tags (ESTs) – short DNA sequences that act as tags for transcribed genes – can be directly related to gene expression. Using ESTs, it will be possible for users to identify the spectrum of variants for a gene, and to follow relations between sequence variation and variation in plant characteristics. Studies of gene function and of the process of gene expression are likely to become increasingly important in providing conservation workers with tools to identify the diversity that is of greatest potential significance for improving production and use of agricultural plants and products.

Pre-breeding, or the transfer of useful traits from exotic germplasm into adapted populations, is an important step in using unadapted germplasm for crop improvement programmes. Molecular mapping, combined with QTL analysis and studies on gene function, can be used to concentrate on valuable new genes in exotic germplasm (Tanksley and McCouch, 1997; Tanksley and Nelson, 1996) and transfer them to improved crop varieties, using conventional crossing combined with marker-assisted selection.

The added resolution afforded by new genetic technologies is generating enormous amounts of data on DNA sequences, molecular maps, metabolic pathways, gene expression profiles and genome structures. This has necessitated the rapid development of new ways to handle such data – a bioinformatics revolution which includes increased computing capacity, new tools for data manipulation and analysis, and new standards for storing, annotating and communicating data. Through the use of the Internet, it is now possible to link information on genome maps, genes, DNA clones, molecular polymorphisms, DNA sequence homologies, links to related information from other Internet resources, the available genetic resources in some of the world's key genebanks and other data as well (for examples see *http://genome.cornell.edu*).

These developments in molecular genetics create new opportunities and new challenges for genebanks. Collections of crop wild relatives will be needed to take advantage of the opportunities offered by molecular methods to locate and transfer their useful variation. Trait-based collections may be developed to bring together materials with specific characteristics (such as abiotic stress tolerance) regardless of biological source. These will provide material for gene transfer, either by traditional plant breeding or by genetic engineering, and for understanding the basis of gene expression. DNA storage is now routine and genebanks need to determine the role that they play in this area. They will also be increasingly required to maintain the well-characterized genetic stock collections that will underpin plant molecular genetic studies. The existence of extensive genome synteny provides a new basis for collaboration between genebanks working on species from the same or related botanical families backing their search for new useful variation.

Perhaps the greatest challenge will be for genebanks to ensure their continuing relevance to germplasm users. As public sector support for breeding work has declined, the opportunities offered by molecular methods to identify useful variation and transfer it to adapted germplasm have dramatically increased. An increased role for genebanks in pre-breeding work seems both necessary and justified.

KEY ISSUES AND IPGRI's CONTRIBUTION

IPGRI has already undertaken a number of activities involving molecular genetics. Research activities have included the development of microsatellite markers for use with coconuts, the molecular characterization of lettuce genebank accessions, the transformation of *Musa* for resistance to black Sigatoka, and the measurement of diversity in forest in India and Costa Rica (Shaanker *et al.*, 1998, 1999; Shaanker and Ganeshaiah, 1997; Cornelius, 2000). An analysis of molecular facilities for genetic resources work in Africa has also been carried out and a number of studies are in progress on the extent and distribution of diversity using molecular markers.

A number of fundamental programme objectives need to be pursued through work developed in at least four different areas: research, capacity building, information technologies and policy. In doing so greater attention ought to be given to the following:

- ensuring that advances in molecular genetics and genomics are made available to the genetic resources community;
- working with genebanks to identify new roles and opportunities that use molecular technologies for maintaining and using diversity;
- seeking to integrate molecular methods and data with other work and with data from other sources. Molecular data do not replace the need for good agromorphological data and for information from social, economic and ecological studies. Often they make such data more important and useful;
- emphasizing useful characters and the development of molecular methods that allow the identification and exploitation of variation at gene loci controlling useful traits;

- ensuring that work on molecular aspects of conservation complements and supports increasing work with farmers and communities and reflects their needs and concerns;
- ensuring that molecular methods are used to support the improved conservation of neglected and underutilized crops. The increased privatization of agricultural research has increased the tendency to focus resources on a few globally important crops, leaving the genetic resources of minor crops increasingly at risk.

Research

Working with appropriate partners, research on topics of global significance that are of highest priority for the genetic resources community is a must. Activities should focus on new opportunities that molecular genetics provides for finding and understanding diversity, managing germplasm and locating useful genes. At the same time, work involving crops identified as needing particular attention (banana and plantain, coconut and neglected and underutilized crops) should receive high priority. Areas of research fall into the two main fields of concentration: conservation and use.

Conservation

Molecular studies of diversity involve the following:

- comparing the diversity patterns obtained with different marker systems (RFLPs, AFLPs, SSRs) and their relation to agromorphological variation;
- determining how molecular diversity distribution is related to geographic, ecological and socioeconomic variation through geospatial analysis;
- measuring the extent of genetic erosion at national, regional or local levels;
- monitoring genetic drift and shift in plant populations conserved *in situ* and supporting on-farm management of agrobiodiversity.

Genebank enhancement It is now possible to use molecular markers to characterize entire collections. Future work in this area will be especially concerned with the problems of managing the large amounts of molecular data generated by these activities and with optimizing characterization procedures.

There are a number of practical aspects of genebank management where using molecular methods can bring substantial benefits. These include identification of duplicates, accession characterization, establishing core collections, identifying major gaps in collections and monitoring genetic changes in *ex situ* collections. In particular, molecular methods will allow more effective monitoring of regeneration and the development of more cost-effective procedures.

DNA sequences are being routinely stored and DNA banks are now beginning to be established. Ways in which gene sequence storage can form an integral part of complementary conservation should be explored. The development of trait-based collections in which the emphasis is on building collections and research programmes around key agronomically important characters (such as abiotic stress resistance) will also be investigated in order to enhance the contribution that genebanks can make to increasing crop production.

Use

Searching for useful traits Research is needed to study the ways in which molecular techniques can be used to facilitate and speed up the identification of plants with specific useful traits. This may well involve an increased role for genebanks in supporting pre-breeding activities through location of useful gene sequences in exotic germplasm and support for marker-aided selection. Important areas of research include the following:

- analysis of the extent of linkage disequilibrium in crop gene pools and hence of the extent to which desirable alleles may be associated with specific markers;
- exploring the use of genome synteny between crops of the same botanical family. This can be used to help identify areas of the genome with specific useful traits in different species. It will be most valuable as a way of using information on a well-studied species, such as rice, to help with conservation and use of genetic resources of less studied species, such as bamboo;
- using core collection and other diversity structuring techniques to improve search procedures in large germplasm collections;
- exploring whether knowledge of specific allele DNA sequences can be used to help locate accessions with those alleles.

Genetic transformation Genetic engineering offers an important way of improving cultivars, not as the only alternative but as the only one once traditional methods cannot provide a sensible solution for a crucial agricultural problem. Such is the case of improving sterile *Musa* cultivars where different genetic transformation methodologies have been developed to obtain *Musa* transgenic plants with increased tolerance to fungal and viral diseases.

Molecular pathology Rapid developments in molecular genetics are presenting significant opportunities for work on plant and germplasm health. Opportunities for developing diagnostic tools to assist safe germplasm exchange in selected crops such as coconuts and *Musa* ought to be monitored.

Capacity Building

One of the biggest challenges for national programmes is to develop facilities and skilled personnel to use the new molecular technologies for conservation and use of plant genetic resources. Capacity should be built by strengthening linkages between those with the relevant skills, equipment and facilities and those involved in conservation. An effective way to do so is the provision of support for strengthening linkages between organizations (such as university departments) with the necessary molecular genetic skills and capacities, and those concerned with conservation within individual countries.

Short training courses on the application of different molecular techniques as well as on-the-job individual training are, then, needed to develop a sufficiently large cadre of genetic resources workers with molecular skills.

Information Management

Molecular genetics studies are dramatically increasing the amount of information available. At the same time, increasing information is becoming available from many other sources. Bioinformatics is a rapidly expanding area that seeks to meet users' needs to gain access to and use such information. The key aim will be to ensure that molecular and other information can be effectively linked to conventional plant genetic resources data and used throughout the world.

Policy

Key concerns will need to be addressed in the policy area. There are already substantial questions and debate about the use of molecular and other biotechnological procedures in agriculture. There are major trade interests as well as biosafety and ethical concerns that affect the use of genetically modified plants. A significant programme of work on genetic resources policy has been developed and will integrate molecular genetic aspects insofar as they impinge on conservation and use of genetic resources. Important areas of concern will include ownership and benefit sharing, and the ways in which gene manipulation affects these; the possible impact of release of novel genes and transgenics into the natural and agricultural environment; and the effect of new technologies and private sector activities on farmer management of crop diversity.

REFERENCES

Ahn, S., J.A. Anderson, M.E. Sorrells and S.D. Tanksley (1993), 'Homoeologous relationships of rice, wheat and maize chromosomes', *Molecular and General Genetics*, 241(5–6), 483–90.

Austin, D.F. and M. Lee (1998), 'Detection of quantitative trait loci for grain yield and yield components in maize across generations in stress and non-stress environments', *Crop Science*, 38(5), 1296–1308.

Bailey, M.A., M.A.R. Mian, T.E. Carter, Jr., D.A. Ashley and H.R. Boerma (1997), 'Pod dehiscence of soybean: identification of quantitative trait loci', *Journal of Heredity*, 88(2), 152–4.

Barker, H. and L. Torrance (1997), 'Importance of Biotechnology for germplasm health and quarantine', in J.A. Callow, B.V. Ford-Lloyd and H.J. Newbury (eds), *Biotechnology and Plant Genetic Resources. Conservation and Use*, Wallingford: Cab International, pp. 235–54.

Butler, D. and P. Pockley (2000), 'Monsanto makes rice genome public', *Nature*, 404, 534.

Cornelius, J. (2000), 'The impact of forest fragmentation on effective population sizes, mating systems and genetic diversity of forest trees in Guanacaste Province, Costa Rica', IPGRI project progress report, IPGRI, Rome.

Devos, K.M. and M.D. Gale (1997), 'Comparative genetics in the grasses', *Plant Molecular Biology*, 35(1–2), 3–15.

Devos, K.M., J. Beales, Y. Nagamura and T. Sasaki (1999), 'Arabidopsis-rice: will colinearity allow gene prediction across the eudicot-monocot divide?', *Genome Research*, 9(9), 825–9.

Dodds, J.H. and L.W. Roberts (eds) (1985), *Experiments in Plant Tissue Culture*, Cambridge: Cambridge University Press.

Dunham, I., A.R. Hunt, J.E. Collins, R. Bruskiewich, D.M. Beare (1999), 'The DNA sequence of human chromosome 22', *Nature*, 402, 489–95.

Engelmann, F. (1997), 'In vitro conservation methods', in J.A. Callow, B.V. Ford-Lloyd and H.J. Newbury (eds), *Biotechnology and Plant Genetic Resources. Conservation and Use*, Wallingford: Cab International, pp. 119–61.

Eshed, Y. and D. Zamir (1995), 'An introgression line population of Lycopersicon pennellii in the cultivated tomato enables the identification and fine mapping of yield-associated QTL', *Genetics*, 141(3), 1147–62.

Foolad, M.R. and F.Q. Chen (1999), 'RFLP mapping of QTLs conferring salt tolerance during the vegetative stage in tomato', *Theoretical and Applied Genetics*, 99(1–2), 235–43.

Foolad, M.R. and R.A. Jones (1993), 'Mapping salt tolerance genes in tomato (Lycopersicon esculentum) using trait-based marker analysis', *Theoretical and Applied Genetics*, 87(1–2), 184–92.

Foolad, M.R., F.Q. Chen and G.Y. Lin (1998), 'RFLP Mapping of QTLs conferring cold tolerance during seed germination in an interspecific cross of tomato', *Molecular Breeding*, 4(6), 519–29.

Grandillo, S., D. Bernacchi, T.M. Fulton, D. Zamir, S.D. Tanksley, G.T. Scarascia-Mugnozza, E. Porceddu and M.A. Pagnotta (1999), 'Advanced backcross QTL analysis: a method for the systematic use of exotic germplasm in the improvement of crop quality', *Proceedings of the XV EUCARPIA Congress. Genetics and breeding for crop quality and resistance. Viterbo, Italy. September 20-25 1998*, Dordrecht: Kluwer Academic Publishers, pp. 283–90.

Lin, X., S. Kaul, S.D. Rounsley, T.P. Shea, M.I. Benito (1999), 'Sequence and analysis of chromosome II of Arabidopsis thaliana', *Nature*, 402, 761–8.

Lindhout, P., S. van Heusden, G. Pet, J.W. van Ooijen, H. Sandbrink, R. Verkerk, R. Vrielink and P. Zabel (1994), 'Perspective of molecular marker assisted breeding for earliness in tomato', *Euphytica*, 79(3), 279–86.

Liu, S.C., S.P. Kowalski, T.H. Lan, K.A. Feldmann and A.H. Paterson (1996), 'Genome-wide high-resolution mapping by recurrent intermating using Arabidopsis thaliana as a model', *Genetics*, 142(1), 247–58.

Mayer, K., C. Schuller, R. Wambutt, G. Murphy, G. Volckaert (1999), 'Sequence and analysis of chromosome 4 of the plant Arabidopsis thaliana', *Nature*, 402, 769–77.

Monforte, A.J., M.J. Asins and E.A. Carbonell (1997), 'Salt tolerance in Lycopersicon species. V. Does genetic variability at quantitative trait loci affect their analysis?', *Theoretical and Applied Genetics*, 95(1–2), 284–93.

Ogihara, Y., K. Isono and A. Saito (1994), 'Toward construction of synteny maps among cereal genomes. I. Molecular characterization of cereal genomes as probed by rice genomic clones', *Japanese Journal of Genetics*, 69(4), 347–60.

Paterson, A.H., Y.R. Lin, Z.K. Li, K.F. Schertz, J.F. Doebley, S.R.M. Pinson, S.C. Liu, J.W. Stansel and J.E. Irvine (1995), 'Convergent domestication of cereal crops by independent mutations at corresponding genetic loci', *Science*, 269(5231), 1714–18.

Pecchioni, N., G. Vale, H. Toubia-Rahme, P. Faccioli, V. Terzi and G. Delogu (1999), 'Barley–Pyrenophora graminea interaction: QTL analysis and gene mapping', *Plant Breeding*, 118(1), 29–35.

Qi, X., R.E. Niks, P. Stam and P. Lindhout (1998), 'Identification of QTLs for partial resistance to leaf rust (Puccinia hordei) in barley', *Theoretical and Applied Genetics*, 96(8), 1205–15.

Ribaut, J.M., C. Jiang, D. Gonzalez de Leon, G.O. Edmeades and D.A. Hoisington (1997), 'Identification of quantitative trait loci under drought conditions in tropical maize. 2. Yield components and marker-assisted selection strategies', *Theoretical and Applied Genetics*, 94(6–7), 887–96.

Shaanker, R.U. and K.N. Ganeshaiah (1997), 'Mapping genetic diversity of *Phyllantus emblica*: Forest genebanks as a new approach for *in situ* conservation of genetic resources', *Current Science*, 73(2), 163–8.

Shaanker, R.U., K.N. Ganeshaiah and K.S. Bawa (1998), '*In situ* conservation of forest genetic resources: a research programme for the Western Ghats, India', IPGRI project progress report, IPGRI, Rome.

Shaanker, R.U., K.N. Ganeshaiah and K.S. Bawa (1999), 'Conservation of Forest Genetic Resources in Western Ghats', IPGRI project progress report, IPGRI, Rome.

Stuber, C.W., S.E. Lincoln, D.W. Wolff, T. Helentjaris and E.S. Lander (1992),

'Identification of genetic factors contributing to heterosis in a hybrid from two elite maize inbred lines using molecular markers', *Genetics*, 132(3), 823–39.

Tanksley, S.D. and S.R. McCouch (1997), 'Seed Banks and Molecular Maps: Unlocking genetic potential from the wild', *Science*, 277, 1063–6.

Tanksley, S.D. and J.C. Nelson (1996), 'Advanced backcross QTL analysis: a method for the simultaneous discovery and transfer of valuable QTLs from unadapted germplasm into elite breeding lines', *Theoretical and Applied Genetics*, 92, 191–302.

Xiao, J.H., J.M. Li, S. Grandillo, S.N. Ahn, L.P. Yuan, S.D. Tanksley and S.R. McCouch (1998), 'Identification of trait-improving quantitative trait loci alleles from a wild rice relative, Oryza rufipogon', *Genetics*, 150(2), 899–909.

10. Biotechnology and traditional breeding in sub-Saharan Africa

Vittorio Santaniello

INTRODUCTION

The popular press has recently echoed some of the criticisms of the biotech industry, employing often contradictory arguments. On the one hand, it has argued that agricultural biotechnologies are bound to benefit only producers and consumers of developed countries. Those, it is claimed, are the only ones that can afford to pay for the sophisticated technologies and the considerable amount of resources needed to promote an active agricultural biotechnology sector. Moreover, the biotech industry being largely financed by private resources, it will only produce those goods and services that can find their way into a rich commercial market, the only one able to repay the investments that are required. This means that biotechnology is biased against the consumers and producers that are poor or that live in developing countries.

At the same time, it is argued, the markets of the developing countries will run the risk of playing the role of 'wastepaper basket' for the outputs of the biotech industry. Consumers in developed countries are increasingly hostile to the products of this industry, which therefore will not find their way into these rich markets. Well-off consumers in developed countries will be able to afford the luxury of 'biologically superior' products. Moreover, producers in developing countries will become increasingly dependent on a proprietary input market, because a new wave of technological innovation will be forced upon them and an increased economic rent will be extracted from their already reduced income.

Again the argument here is that biotechnology is biased against the poor. The former argument, however, is that the poor will suffer because they will be deprived of the benefits of the new biotech products. In the latter, the poor will suffer because they will be forced to consume and produce the new biotech products.

The 'green revolution' has been a determining force that allowed the world to avoid the widespread phenomenon of mass starvation, particularly in Asia. A concentrated effort by scientists and agricultural technicians has made all this possible. At the beginning of the new century we are facing a new challenge. In the next 20 years, 95 per cent of the increase in the world population will be housed in the developing countries. Natural resources and science will have to feed them, without creating undue pressure on the environment.

The task is complicated by the fact the traditional recipes have lost some of their appeal. The traditional tools employed during the green revolution to increase yields and production are now less effective than they were in the 1950s and 1960s. Will biotechnology learn from this experience? What can and should be done to make it possible? Are there some additional lessons we can learn from the most recent performance of science applied to agriculture, that can guide us in the as yet unknown field of biotechnology? These are some of the developmental issues now before us and awaiting an answer.

This chapter tries to make a small contribution in this direction. In particular, it presents an overview of the experience that we have gained from applying traditional breeding to agriculture in Africa and considers whether there is anything that could be relevant to the development of agricultural biotechnology in this region. The first section of the chapter will introduce some of the available information on the present development of agricultural biotechnology in Africa. As will soon be apparent, this sector in Africa is still in its infancy, as only a few countries have started to invest in its development. In the following section we will present a short review of the traditional breeding sector in Africa whose development is a necessary precondition for the creation of a biotechnology sector and indicate many of the issues that need to be solved before a technically and economically sound biotechnology sector can be sustained in Africa. In the final section we will draw a few preliminary conclusions, trying to highlight some of the issues relevant to the development of an agricultural biotech industry in Africa.

BIOTECHNOLOGY IN AFRICA

The information relating to the biotechnology sector in Africa is limited and patchy. No source of information available supplies a systematic survey and a complete coverage of the biotechnology research and production in Africa. The information that will be reported in this section, therefore, is not meant to give a comprehensive picture of the activities in this sector, but more simply is aimed at supplying a view of the types of biotechnology outputs that are produced in this region and to provide an appreciation of the level of technological development that this sector has reached up to now in Africa.

According to FAO there are at present in Africa 20 major research institutions dealing with agricultural biotechnology (FAO, 1996), among them the International Center for Insect Physiology and Ecology, the Kenyan Agricultural Research Institute (KARI) and the International Livestock Research Institute, all in Kenya; the International Institute for Tropical Agriculture (IITA) in Nigeria; the Agricultural Research Council (ARC) in South Africa; the Nikobisson Biotechnology Center at the University of Yaoundé in Cameroon; and the Cocoa Research Institute in Ghana. In addition to these institutions there are minor centres mainly in universities, such as the Jomo Kenyatta University of Agriculture and other academic institutions in Uganda, Ethiopia, Sudan and Burundi (ibid.).

Although the level of activity of the biotech sector in Africa is not very high, there is a scattered set of activities in several areas that can be seen as the seed of a future and larger industry. This activity is the result of pioneering projects often financed by external sources. The fields in which these projects are the production of biofertilizers, the cloning of in vitro plants, bioprospecting of nitrogen fixing species of bacteria and mycorizae, creation of genetic and hybrid variability, bioindustrial production of plant metabolites of medical significance (*Rauflora serpentina*, *Abrus precatorious*, *Terhosia Vogelli*) and development and release in South Africa and Nigeria of commercial transgenic maize and cotton. The appendix to this chapter reports a list of African countries and the field of biotech activities in which their research institutions operate (Brink *et al.*, 1998).

With few exceptions, all of the above institutions deal with what can be identified as the first generation of biotechnology techniques. Plant tissue culture or clonal propagation are used to reproduce vegetatively virus-free material for export crops such as coffee, cocoa, oil palm and tobacco (Gbewonyo, 1997) and food crops such as cassava, sweet potatoes, yams, potatoes, bananas and plantain (Woodward *et al.*, 1999; Brink *et al.*, 1998). These techniques are used for mass production of disease-free plants as well as regeneration systems for plant transformation. By concentrating on tissue culture, these laboratories have developed important skills for the management of biotech laboratories, creating a very valuable asset for further development of this sector in Africa.

The production of disease-free plant material has been practised commercially in Kenya since the middle of the 1980s, and since 1994 banana plantlets are sold commercially by a private company (Qaim, 1999). IDRC Canada has financed a development project that will pilot test the process of sourcing selected banana planting materials, multiplying them by tissue culture at the Jomo Kenyatta University of Agriculture and adapting, promoting and distributing the final planting material to selected farmers in three banana-growing areas of Kenya. The objective of the project is to assess the feasibility, cost effectiveness and potential benefits of conducting such an enterprise on a commercial basis under Kenyan conditions (R. Diprose, personal communication).

The production of biotechnology products is a necessary but not sufficient condition to generate and disseminate its potential benefits. These goods need to reach the final users (farmers and consumers), who can benefit respectively from increased yields and lower food prices. This flow can either be achieved through the channel of the private seed industry or by employing public sector means such as the extension service. In most African countries several factors have conspired against the creation of a well functioning seed industry, among them the limited size of a solvent market, a deficient IPR legislation, and the fact that in several instances the seed industry has been confined to an ineffective public sector. This justifies the concern of those who say that, although biotech products can be obtained in Africa, their channelling to final consumers will be lacking and their economic impact therefore limited.

The presence of a well established traditional breeding programme has facilitated in a few institutions molecular markers application (diagnostic, fingerprinting and marker assisted breeding). The capacity to produce transgenic plants has been limited to South Africa and Nigeria.

A serious constraint on the development of the biotech sector in Africa has been the limited number of highly skilled scientists available. It has been often complained that this already limited number is made even more scanty by the reluctance of many African scientists, trained overseas, to come back to their home countries to earn a meagre salary, to work in a suboptimal research environment with inadequate infrastructure and limited resources to run research projects, isolated from the international scientific community and with the consequent difficulty of keeping up with scientific development in their fields. The recent development in communication technologies might make this sense of isolation and the danger of a rapid obsolescence of the human capital less of a problem. Obviously, the other constraining factors remain.

What is not always well recognized, however, is that there also exists an internal brain drain, which sees a number of Western-trained scientists, within a few years of returning home, being attracted by working opportunities not necessarily related to research. From a more general point of view, this internal migration might not necessarily imply a social loss, but inevitably it acts as a limiting factor to the development of a science sector so heavily dependent on human capital.

Biotech research activities in Africa are thus hampered by lack of adequate infrastructure, research equipment and facilities, and skilled personnel (Villalobos, 1995). In some cases investments in human capital, infrastructure and operating capital have not been well balanced, and this has created a problem of its own. In Kenya there has been a strong emphasis on producing biotech scientists, but their numbers have increased faster than the resources devoted to this type of research. The financial resources per scientist have in fact decreased. Between 1989 and 1996, annual expenditures in agricultural biotech went from US$ 2.5 million to US$ 3.0 million. The number of scientists, however, increased at a faster rate and this has caused a decrease in financial resources per research project that went from US$ 77.2 thousand in 1989 to US$ 45.5 in 1996, with an average decrease of 7.2 per cent per year (Falconi, 1999).

THE TRADITIONAL BREEDING SECTOR IN AFRICA

What can we gain from an analysis of the traditional breeding sector that can be useful for forecasting the most likely pattern of development of the agricultural biotechnology sector?

1. Owing to the present stage of growth of the agricultural biotechnology sector in Africa, any analysis of this sector can only be a sort of ex ante assessment of the possible issues, products and potentials of its development. An analysis of the experience in the traditional breeding sector is part of this ex ante exercise. The experience gained in the traditional breeding makes it possible to simulate the possible effects of a more intensive use of science in agriculture.
2. Agricultural biotechnology and traditional breeding supply the same market with products that are either complementary or substitutes. Biotech and traditional breeding are to a large extent part of the same industry and they will face an identical set of structural developmental issues. The same bottlenecks that are constraining the development of the traditional breeding sector will be found relevant for agricultural biotechnology as well.
3. The existence of a well established national traditional breeding activity in a country is a necessary precondition to envisioning and implementing an effective agricultural biotechnology programme. As in many other instances, development cannot jump and so it is not realistic to think about a biotechnology sector by jumping over a proper development of the traditional breeding sector. The stage of development of the traditional breeding sector can tell us how far we are from

a full development of a biotech industry and what are the steps we still need to climb.

4. The development profile of traditional breeding can prefigure the most likely growth of the agricultural biotechnology sector.
5. In particular, the relationships between national research centres (NARs) and the international research centres (IARCs) should be closely scrutinized, because they present an interesting pattern on which to model the role that the latter can perform to promote biotech development in Africa. Past experience with IARCs in Africa has proved that NARs have benefited considerably, directly and indirectly, from the research and experimental work done by IARCs, and that their achievements increased significantly after the IARCs diverted resources to the study of breeding and crop management issues in Africa. For the same reason, the hesitancy that the IARCs are showing in undertaking a more active role in the development of agricultural biotechnology is a source of concern.

The Production of Improved Varieties

All countries in the developing world have benefited from the yield increase made possible by the introduction of improved varieties. The profile and timing of this increase have been different depending upon crops and regions. Production increase for maize and wheat tapered off in the early 1980s. The same thing happened with other crops towards the end of the same decade. In Africa, which was a latecomer, contrary to the present experience of Asia and Latin America, yields are still in the upward trend (Evenson, 2000).

Regardless of the profile and timing of the yield increases, genetic research and traditional breeding have proved to be well suited tools to achieve a forward shift in the yield frontier, to pursue an increase in plant resistance to biotic and abiotic stresses, to stabilize production and, finally, to attain an improved input–output efficiency. To be effective, however, this instrument requires the availability of quality germplasm. Evidence shows that a forward leap in production has resulted where scientists have had a wide access to a wealth of genetic resources (Pingali, 1999). This has been proved to be repeatedly true for Asia and Latin America. In Africa, contrary to the Asian and Latin American experience, the production and diffusion of improved varieties has been hampered by the scarcity of collected and properly evaluated germplasm.

Until recently, genetic improvement in Africa was heavily (and in several instances still is) dependent on research and germplasm collected and evaluated outside the region.[1] The original mandate of the West Africa Rice

Development Association (WARDA) (Dalton and Guei, 1999), for example, was focused on adaptive field trials of exotic imported material. Only in the second half of the 1990s did WARDA depart from this original strategy and start emphasizing the need to develop an indigenous varietal improvement strategy.

This dependence on Asian genetic materials and varieties impeded a wider dissemination of improved rice varieties in Africa. Although imported varieties were tested to choose those that were better adapted to local conditions, their origin inevitably decreased the available degrees of freedom. The criteria that led to the selection of the popular IR8 rice variety, for example, demonstrate well the nature and implications of the links that exist between the breeding objectives and the relevance of the research programme to a specific environment.

All through the years rice producers in Asia had selected rice varieties that were responsive to photoperiod. These varieties flowered at the peak of the rainy season, avoiding stress during the reproductive phase that, for rice, is the most sensitive period. In the breeding programmes geneticists were induced to use China-derived varieties to take advantage, among other things, of their unresponsiveness to photoperiod. This new trait, together with the shorter and fixed growth period, allowed farmers around the world – that had access to irrigation and appropriate climatic conditions – to move to a multiple cropping system.

These characteristics, however, made varieties unsuitable for rainfed African agriculture, which therefore still had to depend on traditional varieties. To have depended on a variety with less 'buffering capacity' than the traditional photoperiod types, could have been very risky because critical growing periods could have coincided with periods of drought or of shortage of labour.

Most of the improved Asian rice varieties were dwarfed or semi-dwarfed, which made it possible to improve the grain–straw ratio and the efficient use of fertilizers, and helped the plant to sustain the weight of a heavy grain panicle. In contrast, the number of deep water improved rice varieties produced by the national and international research centres has been minimal. In areas where the control of water was less than optimal, however, drought stress might have reduced height even more than the desired level, making plants suffer and decreasing their productive capacity. Varieties often typically found, and needed in the African environment instead, were of the floating or deep water type because of the difficulties of controlling the water flow in the farmers' fields. Dwarfed rice, therefore, was not an optimal choice where the water flow was not under farmers' control. Again this varietal development, which did not take proper account of this environmental constraint, was of limited value for African agriculture.

Adoption of New Improved Varieties

The release of improved varieties, however, is a necessary but not sufficient condition. To generate the desired impact on producers' and consumers' welfare, the new improved genetic material has to be adopted by farmers. Several factors have influenced the adoption rate of improved varieties by African farmers. Those adoption rates have historically been different according to species, varieties, geographical locations and social groups.

Generally speaking, the experience of at least the last two or three decades in Africa has been that, when improved varieties have become available and have been positively evaluated by farmers, they tended to be widely adopted in a relatively short time. Progress in the adoption of improved maize varieties has been almost as rapid in sub-Saharan Africa as in Asia and Latin America (Byerlee and Heisey, 1996). In a study of the adoption rate of two new bush bean varieties, K131 and K132, by Ugandan farmers it has been found that it took only four years before 74 per cent of the studied area had been covered by the new breeds (David *et al.*, 1999). In 1993, in Sierra Leone, 90 per cent of lowland farmers were using the improved rice variety ROK14; that is, only five years after the new variety had been released (Edwin and Master, 1998). Between 1986 and 1990, in the mangrove areas of Sierra Leone, the percentage of farmers that moved from existing to improved rice varieties went from 16 to 56 per cent (Adesina and Zinnah, 1993).

Genetically improved varieties have the advantage of being an embodied technology and its adoption usually does not necessarily require major changes in crop management practices. Moreover, seed technology is simple to evaluate and this favours the conversion of information awareness into knowledge and adoption (Evenson, 1997). The ease with which new varieties can be evaluated and the relatively minor changes that their adoption imposes on crop management help to explain the rapidity with which this technological innovation disseminates among farmers.

Impact studies of innovations tend to show that production and adoption of new varieties produce more enduring effects than other technical improvements. The rate of adoption of an innovation depends upon the marginal rate of return of the cash outlay needed for its implementation. This in turn is a function of the input–output price ratio. Better crop management, like a more intensive use of fertilizers, can bring significantly larger yields, even at levels comparable with those generated by improved varieties. However, the influence of these potential improvements on farmers' behaviour seems to be rather different from the effects of the adoption of improved varieties.

In the Ghana maize project the shift from local to improved varieties did cause a production increase of 88–102 per cent, while the addition of fertilizer to traditional varieties had permitted a production increase of 81 per cent. The production increases that could be obtained through the use of fertilizer, however, tended to produce less lasting effects on farmers' choices than those generated by improved seeds. One-third of farmers that started using fertilizers in maize production discontinued this technology after they had been using it for some time, while less than 10 per cent of farmers that had turned to improved varieties discontinued them (Morris *et al.*, 1999).

Often the adoption of improved varieties acts as a catalyst for other improved practices such as fertilization, row planting and so on. The catalyst effect, however, is not always the rule, because improved varieties tend to perform better then traditional varieties even under non-optimal conditions. In Nigeria, 62 per cent of soybean growers had chosen improved varieties because they outperformed traditional varieties in poor soil conditions and with no fertilizer addition (Sanginga *et al.*, 1999). In Ghana, 40 per cent of Ashanti farmers were using improved rice varieties, but a much smaller percentage of them were also employing other improved inputs and complementary crop management techniques (Dankyi *et al.*, 1996).

For a technology to be widely adopted, the existence is necessary of a set of complementary conditions, such as an efficient extension and input distribution system, and an appropriate economic environment able to generate proper economic incentives. Those conditions, however, tend to play a different role depending upon the crop and the seed production technology. Usually vital to the dissemination of a new technology is the role of the agricultural extension service. Often extension officers are the first suppliers of the new seeds, as they promote the creation of demonstration plots to show farmers the newly introduced traits and advantages that could be derived from the adoption of the new varieties. Depending on the type of crop and of the market structure, private seed breeders and producers replace public extension.

In Ghana, over 50 per cent of the farmers that had adopted improved maize varieties had obtained them from extension officers, although the role of private input dealers was increasing (Morris *et al.*, 1999). For soybean in Nigeria (Sanginga *et al.*, 1999), market proximity, farmer's age, gender, number of contacts with extensions, yields and time of maturity were the variables that were found to influence the decision to adopt the improved varieties. In this case, however, contacts with extension officers seemed to play a less determinant role than for maize. Only 22 per cent of farmers in fact had received improved soybean seeds from extension officer, while a greater role in the dissemination of the improved genetic material was played by the hand-to-hand transfer of seed among farmers.

An important issue that relates to the impact of modern varieties is whether there exists a gender bias in the access to improved varieties or, on the contrary, this technological innovation is gender-neutral. Some evidence leads to the conclusion that, although indirectly, men are favoured in this race. Men in fact tend on average to be favoured because they are better educated, have more frequent contact with extension officers and have easier access to financial resources than women.

An impact study of the adoption rate of new soybean varieties in Nigeria (Sanginga *et al.*, 1999) has found that men farmers were the first to adopt the new seed, while women farmers were significantly involved only four years after those seeds had been officially introduced. In the following seven years, men farmers always had a lead in the adoption of new varieties. In the last year for which the data were available, 75 per cent of men farmers were growing the improved varieties of soybean, while only 62 per cent of the women farmers were adopting the new technology.

In some cases market structure has favoured a selection among alternative seed production techniques. In the case of maize, market structure and other socioeconomic factors have favoured a choice of a technically second-best solution that is, however, better from the economic standpoint.

It has been estimated that, at the beginning of the 1990s, approximately 43 per cent of the maize area in Africa was covered by improved germplasm. In a few cases, such as Kenya, Zimbabwe, Zambia, Lesotho and Swaziland, where there is a relatively well functioning national seed industry, hybrids of maize have had the largest share of the improved areas. In all other cases, farmers have given their preference to open pollinate varieties (OPV). OPV, although less productive than hybrids, have the advantage that seed can be saved by farmers with no major loss of genetic purity and therefore yields. The same is not true for hybrids, where heterosis is mainly limited to F1. Yields with improved maize varieties, in Africa, have been significantly higher than traditional varieties. Hybrids in good farming conditions produce at least 40 per cent more than existing local varieties. In less favoured conditions, such as in drought problem areas, yield increase might be slightly less, around 30 per cent. Yield increases of OPV are generally lower, although still significant, as they average around 14–25 per cent above unimproved material (Byerlee and Heisey, 1996).

OPV are better tuned, however, to the needs of smaller farmers because they make them less dependent on the input markets, and perform better then hybrids in the absence of a well functioning seed market. Maize seed production in Ghana was the responsibility of a government-owned company which was plagued by a chronic shortage of funds and of trained personnel, and was unable to supply farmers with quality seed in a timely fashion. In this situation to have depended upon the national seed supply would have been a risky decision, and turning to OPV was the only available solution.

The absence of a well established and enforced intellectual property right (IPR) system has contributed to depriving Africa of a well functioning commercial market for improved varieties and has forced the public sector to assume a leading role in the production and distribution of genetically improved material also for those crops where, in other regions, the private sector plays a leading role.

At present, maize research in Asia is done almost entirely by the private sector. In Africa, on the contrary, the public sector is heavily involved in the genetic research for this crop. The different role and involvement played by the private and public sector in the two regions is explained by the diversity in the development and implementation of the IPR legislation. In Asia, where IPR legislation has created a market for maize seed, the private sector is involved in research for this crop. In Africa, on the other hand, where IPR legislation is either non-existent or not effectively implemented, the public sector plays a pre-eminent role in this area.

All of the above, together with the weakness of the extension service, has not favoured the establishment of an effective system of information flow, on the performance of the experimented technologies and on farmers' needs, between the final users and the researchers. This lack of flow of information has increased transaction costs and contributed to the inefficiency in national agricultural research. To reduce these costs, the public sector research should adopt a strategy of extensive testing of improved varieties at the farm level. This would foster farmers' participation in improving the technologies and in making them more relevant to the needs of final users. Moreover, by operating at the farm level, scientists would get a better appreciation of other factors that, although not directly related to the technology being tested, could operate as limiting factors to its deployment. In addition this acquired knowledge would help improve the design of new crop management technologies.

Traditionally, the African national research systems have been plagued by lack of infrastructures and of operating funds. Despite their operational difficulties, however, NARs in Africa have been able to make a valuable contribution to varietal production. Between 1980 and 2000 NARs in West Africa have released on average eight rice varieties each year. By 2004, an additional 122 varieties will be made available for farmers to test and eventually to adopt. In this effort the role of the CGIAR has been significant either in providing germplasm or in supplying crosses and advanced breeding lines.

Adoption of these new varieties varies according to the ecology of the area. It is highest in irrigated lowland and lowest in upland rice. The relation between the adoption rate of the new varieties and the ecological condition finds its justification in the crop management tradition for rice in the region. Rice production in irrigated lowland is a relatively new development in Africa, therefore varieties and crop management techniques have been heavily borrowed from the plentiful Asian supply. Local experience and local genetic material readily adaptable to these conditions were not available. It has been relatively easy, on the one hand, to find appropriate material, but, on the other hand, lack of local material has lowered the *barrier to entry*. Upland rice in contrast has a long tradition in Africa and therefore farmers, through time, have selected a large number of varieties well adapted to a territory with a considerable location variance (Dalton and Guei, 1999).

SUMMARY AND CONCLUSIONS

In an economic system that is progressively becoming more knowledge-intensive, science and therefore biotechnology are bound to play an increasing role in the agricultural sector of the developed as well as of the developing countries.

Information on the current status of the agricultural biotechnology in Africa is limited. What we do know, however, is sufficient to conclude that, although this sector is still in its infancy, the overall situation is not static and there is a good number of activities going on. Similarly to what has been the experience with the traditional breeding programme, Africa's biotech sector started later, and it will take some time before it will gain momentum.

A large share of biotech activities are located in the national and international public sector or are financed by external sources. This is a source of strength but also of weakness. Those programmes enjoy technical assistance from outside but are heavily dependent on the sources of financing over which they have a rather minor control. All these projects are generating expertise that is vital for the future development of the biotech sector in Africa.

Most of the institutions involved in biotech in the region operate on what can be defined as first-generation biotech, such as tissue culture and cloning. Where there has been a well established traditional programme the biotech sector has been able to venture into more advanced stages such as marker-assisted breeding. Only South Africa and Nigeria have been able to go as far as producing transgenic varieties. The traditional breeding sector plays a pivotal role in the development of the biotech sector. It supplies the same market, and its operating experience and management problems mirror many of the potential bottlenecks that the development of the biotech sector will experience in Africa (lack of communication with the final users, weakness of the extension services, lack of infrastructures, operating funds and skilled personnel, and so on).

All countries have benefited from genetic research: Asia and Latin America first and Africa at a later stage. In the first two regions, production increases are still on the agenda but at a much lower intensity than in the past. Africa, in contrast, is still experiencing a period of relatively high production growth.

What has made Africa come to biotechnology later? Primarily the scarcity of a properly collected and evaluated germplasm and the lack of a sufficiently autonomous traditional breeding capacity. This has made this region dependent on the outside world in terms of improved genetic material. Although new varieties were tested and selected accordingly to African needs, that was not enough because often what was needed just was not available, nor was the demand strong enough to generate an appropriate response. This is a lesson that traditional breeding can pass on to the gene revolution. Africa needs her own research structure to deal with her own specific problems. Pests, diseases, the mix of natural resources, social and economic environments are sufficiently specific to require a proper approach.

The production of improved material and the degree of involvement of the public and of the private sectors in the development of the biotech industry will depend upon the extent to which a proper economic environment, conducive to private investments, will be favoured. Here the primary role will be played by the effective enactment of an IPR legislation.

Whenever African farmers have had access to improved varieties, well adapted to the local conditions, they have adopted them with the same speed as farmers in other parts of the developing world. Almost every time the adoption rate of new varieties had been lagging behind this has been so because they were either not suited to the local ecological characteristics or were at odds with the social and economic conditions. Farmers in Africa are no more progressive or conservative than farmers in other parts of the world, but the conditions in which they operate are different and often more complex than elsewhere.

The transfer of new technology through seed is easier to accomplish and can be done more efficiently than with other types of technological innovations. However, owing to the weakness of the input market structure and of the public extension service, to be really effective new varieties will have to embody as much as they can. Pest and drought resistance and capacity to mobilize nutrients like phosphorus and nitrogen, are some of the traits that biotech can deliver and that would be most useful to African farmers because they will lower their dependency upon an input market that still remains largely inefficient.

Finally, the cooperation between NARS and IARC will be a vital element for the development of a biotech sector in Africa. This cooperation in the traditional breeding sector has proved to be highly productive. It is for this reason that it is hoped that the IARC will soon overcome the doubts that they currently seem to have on operating a more active biotech research programme.

NOTE

1. There are some noticeable exceptions. In the case of rice, for example, the National Cereals Research Institute in Nigeria and the Rokupr Rice Research in Sierra Leone have been working on varietal improvement in Africa for over half a century. The overall research capacity in plant breeding and genetic resource conservation in this area, however, is limited mainly owing to limited operational funds and weak capital infrastructure. Only a few NARS conduct large-scale breeding programmes and an even smaller number of them have a genebank (Dalton and Guei, 1999).

REFERENCES

Adesina, A.A. and M.M. Zinnah (1993), 'Technology Characteristics, Farmers' Perception and Adoption Decisions: A Tobin Model Application in Sierra Leone', *Agricultural Economics*, 9(4).

Brink, J.A., B.R. Woodward and E.J. DaSilva (1998), 'Plant Biotechnology: a Tool for Development in Africa', *EJB Electronic Journal of Biotechnology*, 1(3), December.

Byerlee, D. and P.W. Heisey (1996), 'Past and Potential Impact of Maize Research in sub-Saharan Africa: a Critical Assessment', *Food Policy*, 21(3).

Dalton, T.J. and R.G. Guei (1999), 'Ecological Diversity and Rice Varietal Improvement in West Africa', a report submitted to the GCIAR impact assessment and evaluation group, germplasm impact study.

Dankyi, A.A., V.M. Anchirinah and A.O. Apau (1996), 'Adoption of Rice Technologies in the Wetlands of the Ashanti Region of Ghana' (mimeo), Rice Economics Task Force of West Africa.

David, S., R. Kirkby and S. Kasozi (1999), 'Assessing the Impact of Bush Bean Varieties on Poverty Reduction in Sub Saharan Africa: Evidence from Uganda' (mimeo), CIAT, Pan African Research Alliance.

Edwin, J. and W.A. Master (1998), 'Returns to Rice Technology Development in Sierra Leone', *Sahelian Studies and Research*.

Evenson, R.E. (1997), 'Extension, Technology and Efficiency in Agriculture in Sub-Saharan Africa', The Economic Growth Center, Yale University, Center Paper no. 518.

Evenson, R.E. (2000), 'Crop Genetic Improvement and Agricultural Development' (mimeo).

Falconi, C.A (1999), 'Agricultural Biotechnology Research Indicators and Managerial Considerations in Four Developing Countries', in J.I. Cohen (ed.), *Managing Agricultural Biotechnology – Addressing Research Program Needs and Policy Implications*, Dordrecht: Kluwer.

FAO (1996), *The State of the World's Plant Genetic Resources for Food and Agriculture*, Rome: FAO.

Gbewonyo, K. (1997), 'The Case for Commercial Biotechnology in Sub Saharan Africa', *Nature Biotechnology*, 15, April.

ISAAA (1997), *ISAAA Annual report 1996. Advancing Altruism in Africa*, Ithaca, NY: ISAAA.

Morris, M.L., R. Tipp and A.A. Dankyi (1999), 'Adoption and Impact of Improved Maize Production Technology: A Case Study of the Ghana Development Project', Economics Program Paper 99-01, CIMMYT, Mexico.

Pingali, P.L. (ed.) (1999), *CIMMYT 1998–99 World Wheat Facts and Trends. Global Wheat Research in a Changing World. Challenges and Achievements*, Mexico: CIMMYT.

Qaim, M. (1999), *Assessing the Impact of Banana Biotechnology in Kenya*, Bonn: ZEF.

Sanginga, P.C., A.A. Adesina, V.M. Manyong, O. Otite and K.E. Dashiell (1999), 'Social Impact of Soybean in Nigeria's Southern Guinea Savanna', International Institute of Tropical Agriculture, Ibadan, Nigeria.

Villalobos, V.M. (1995), 'Biotechnology in agriculture: how to obtain its benefits while limiting risk', in *Induced mutations and molecular techniques for crop improvement*, Vienna: IAEA.

Woodward, B., Johan Brink and D. Berger (1999), 'Can Agricultural Biotechnology Make a Difference in Africa?', *AgBioforum*, 2(3&4).

APPENDIX

Table 10A.1 Africa, activities in biotechnology by country

Country	Biotechnology area
Burkina-Faso	Biological nitrogen fixation, production of legumes inoculants, fermented foods, medicinal plants
Burundi	In vitro production of ornamental plants: orchis, tissue culture of medicinal plants, micropropagation of potato, banana, cassava and yam. Supply of disease-free in vitro plants.
Cameroon	Plant tissue culture of *Theobroma cacao* (cocoa tree), *Dioscorrea spp* (yam) and *Xanthosoma mafutta* (cocoyam) Use of in vitro culture for propagation of banana, oil palm, pineapple, cotton and tea
Congo	In vitro culture of spinach (*Basella alba*) Bioprospecting of nitrogen fixing species
Congo, Democratic Republic	In vitro propagation of potato, soybean, maize, rice and multipurpose trees (*Acacia auriculifotius* and *Leucaena leucocefhala*) Experimental production of rhizobial-based biofertilizers Tissue culture of medical plants (*Nuclea latifolia* and *Phyllanthus niruroides*)
Côte d'Ivoire	In vitro production of coconut palm (*cocos nucifera*) and yam Virus-free micropropagation of egg-plant (*Solanum spp*) Production of rhizobial-based biofertilization
Ethiopia	Tissue culture research applied to tef Micropropagation of forest trees
Gabon	Large-scale production of virus-free banana, plaintain and cassava plantlets
Ghana	Micropropagation of cassava plantlets (*Manihot esculenta*), banana/plantain (*Musa spp*), yam, pineapple and cocoa Polymerase chain reaction (PCR) for virus diagnostic
Kenya	Production of disease-free plants and micropropagation of pyrethrum, banana, potato, strawberries, sweet potato, citrus fruits, sugar cane Micropropagation of ornamentals (carnation, alstroemeria, gerbera, anthurium, leopard orchids) and forest trees In vitro selection of salt tolerance in finger millet Transformation of tobacco, tomato and beans Transformation of sweet potato with proteinase inhibitor gene Transformation of sweet potato with Feathery Mottle Virus, Coat protein gene (Monsanto, ISAAA, USAID, ABSP, KARI) Tissue culture regeneration of papaya In vitro long-term storage of potato and sweet potato Marker-assisted selection in maize for drought tolerance and insect resistance
Madagascar	Tissue culture supporting conventional production of disease-free rice and maize plantlets, and medicinal plants Production of biofertilizers to boost production of groundnut (*Arachis hypogea*), bambara groundnut (*Vigna subterranea*)

Table 10A.1 (continued)

Country	Biotechnology area
Malawi	Micropropagation of banana, trees (*Uapaca*) tropical woody species, tea
Nigeria	Micropropagation of cassava, yam and banana, ginger Long-term conservation of cassava, yam and banana, and medicinal plants Embryo rescue for yam Transformation and regeneration of cowpea, yam, cassava and banana Genetic engineering for cowpea for virus and insect resistance DNA fingerprinting of cassava, yams, banana, pests and microbial pathogens Genome linkage maps for cassava, yams and banana
Ruanda	Production of rhizobial-based biofertilizers and *Azolla* for rice cultivation Tissue culture of medical plants and micropropagations of disease-free potato, banana and cassava
Senegal	Production of rhizobial and mycorhizal-based biofertilizers for rural markets In vitro propagation of *Faidhebia albida*, *Eucalyptus canaldulensis*, *Sesbania rostrae*, *Acacia senegal*
South Africa	Genetic engineering: cereals (maize, wheat, barley, sorghum, millet, soybean, lupines, sunflower, sugarcane); vegetables and ornamentals (potato, tomato, cucurbits, ornamental bulbs, cassava and sweet potato); fruits (apricots, strawberry, peach, apple, table grapes, banana) Molecular marker applications: diagnostic for pathogen detection, cultivar identification (potatoes, sweet potato, ornamentals, cereals, cassava); seed lot purity testing – cereals; marker-assisted selection in maize, tomato; markets for disease resistance in wheat, forestry crops Tissue culture: production of disease-free plants (potato, sweet potato, cassava, dry beans, banana, ornamental bulbs); micropropagation (potato, ornamental bulbs, rose rootstocks, chrysanthemum, strawberry, apple rootstocks, coffee, banana, avocado, date palm); embryo rescue of table grapes, sunflower and dry beans; in vitro selection for disease resistance (tomato nematodes, guava wilting disease; long-term storage (potato, sweet potato, cassava, ornamental bulbs); in vitro genebank collection (potato, sweet potato, cassava, ornamentals); forest trees, medicinal plants, indigenous ornamental plants
Uganda	Micropropagation of banana, coffee, cassava, citrus, gnanadella, pineapple, sweet potato and potato In vitro screening for disease resistance in banana Production of disease-free plants of potato, sweet potato and banana
Zambia	Micropropagation of cassava, potato, trees (Uapaca), banana Host Nordic-funded genebank of plant genetic resources
Zimbabwe	Genetic engineering of maize, sorghum and tobacco Micropropagation of potato, cassava, tobacco, sweet potato, ornamental plants, coffee Marker-assisted selection

Source: Brink *et al.* (1998).

Conclusion

11. Policy options for the biotechnology revolution: what can be done to address the distributional implications of biotechnologies?

Timothy Swanson and Timo Goeschl

THE IMPACTS OF GURTs

Genetic use restriction technologies (GURTs) is a generic term that describes a number of different gene-manipulation techniques that can introduce a 'genetic switch' into seed-producing organisms.[1] This switch mechanism may be used to prevent the unauthorized use of the seed itself or of particular traits inherent in the seed. This technology has been dubbed 'terminator genes' by a Canadian NGO, the Rural Advancement Foundation International (RAFI). This term is now widely used in the popular press. The clear purpose of this technology is to enhance the appropriation of the values of innovation from the R&D process. The extent of concrete and specific information is limited by the fact that no real-life example of such a mechanism is known to exist. Nevertheless, there can be no doubt about the practical as well as theoretical feasibility of the technology (Jefferson *et al.*, 1999).

In their currently proposed form, GURTs affect the reproductive capacity of the whole plant. The technology protecting both the 'hardware' and the 'software' of the variety has been termed 'variety-based GURT' or 'V-GURT'. V-GURTs have been the primary objective of research and development. A technological variation that selectively affects the reproduction of particular traits in the next crop generation ('trait-based GURT' or 'T-GURT') has been proposed, but is at the moment only at a putative stage (Crouch, 1998).

GURTs represent a generic technology whose mechanism applies to *all seed-producing organisms*. This means that, in theory, no principal impediments exist for extending this type of biotechnology to other agricultural areas such as fisheries or livestock (Jones, personal communication). There are, however, commercial reasons that are likely to confine the application of GURTs to annual crops and will prevent their extension into other areas of food production. The commercial suitability of applying a technology that enforces seed repurchase every season to annual crops lies in the fact that the planting and fruiting cycle recurs annually. This relatively rapid rate of growth enables several planting/harvest cycles to occur during the course of a single R&D (research and development) cycle. This means that the number of applications for each innovation and therefore of seed repurchase is relatively frequent and therefore commercially meaningful. For other commercially used organisms, such as livestock, fish and trees, the rate of turnover is significantly lower since the life cycle of the R&D embodying organism is long relative to the R&D cycle. This makes the application of GURTs commercially meaningless, because it is intended to enforce seed repurchase between innovations. In short, GURTs are commercially limited to the annual crop species, such as wheat, rice and soybeans.

Therefore GURTs should be seen primarily as a technology for inducing annual seed repurchases. This is intended to enhance appropriation by seed companies of the value of innovations in agricultural R&D. As a by-product, GURTs will also introduce significant changes in the overall system by which agricultural R&D is undertaken. Since the innovative characteristics will be introduced within plant lines that are not interbreedable, the only sector that will be able to continue to undertake R&D down these plant lines will be the private one that introduces the characteristics. Therefore GURTs may be viewed as a new R&D system that will be substituted more or less in whole for the existing set of complex and diverse systems. This substitution will have significant implications for both the current practitioners of agricultural R&D (farmers, the public sector) and the way in which lands are used for these purposes.

The term 'agricultural R&D' is intended very broadly here. It refers to the various methods by which pest and disease problems might be solved in agriculture. These range from traditional farmer use of diverse plant varieties, and the farmer's observation and breeding of useful traits and characteristics, to private plant breeder's use of stored plant varieties, and incorporation into existing lines. Clearly, the different ways in which agricultural problems are solved (the R&D system) has important implications for the rate at which productivity increases in agriculture, the distribution of these gains and the way in which lands are used in achieving these gains.

The first interpretation sets GURTs into the context of existing agricultural R&D systems and assesses the impacts of a new system becoming available. These impacts are transmitted along two lines: (a) appropriation of rents from R&D (see Fisher, this volume), and (b) productivity increases (see Traxler *et al.*, 1995; Huffman and Evenson, 1993; Schmidt 1984). The R&D of crop improvement is generally regarded as the major source of increases in productivity. The structure of agricultural R&D therefore counts among the most important socioeconomic effects of the arrival of new agricultural technologies (for a related study on BT cotton, see Falck-Zepeda and Traxler, 1998) and therefore also of GURTs.

The second interpretation sets GURTs in the context of existing patterns of land use and assesses the impacts of a new system being adopted. The changes in land use induced by GURTs can be classified into two distinct effects: (a) expansion of agriculture, and (b) intensification of agriculture. These changes in the allocation of lands to agricultural uses have been identified as two major causes of loss of environmental amenities, especially the loss of biodiversity in nature and agriculture (Southgate, 1997; Swanson, 1995). Land use changes represent one of the most important environmental dimensions of the adoption of GURTs (see Chapters 9 and 10 in this volume).

How will GURTs change the structure of agricultural R&D and the patterns of land use? To answer this question, information about the current system is required. Table 11.1 presents the dimensions along which GURTs will be assessed for their socioeconomic and environmental impacts. These dimensions, share of R&D between different contributors and land use patterns, are identified for the currently existing systems in agriculture. At the moment, three stereotypical systems can be identified (cf. Srivastava *et al.*, 1996, for a similar typology of agricultural systems):

1. 'traditional agriculture', characterized by a low volume of rent creation and appropriation in R&D, and the predominance of farmer-based plant breeding as a source of crop improvement;
2. 'high-yielding agriculture', characterized by a high volume of rent creation in R&D, and predominant importance of public plant breeding due to insufficient scope for appropriation; and
3. 'F1 hybrid-based agriculture', characterized by a high volume of rent creation and appropriation in R&D, and the predominance of private plant breeding over public or traditional farming systems.

Each of these three systems may be *stereotyped* in a manner to indicate both the way in which it distributes the rents from agricultural R&D, and the way in which it allocates land use. Table 11.1 demonstrates these stereotypes.[2] The figures in this table illustrate how the various systems allocate R&D 'rents' (benefits) differently, and how land use is allocated differently. For example, the table typifies 'traditional agricultural systems' as those in which public breeders generate and appropriate 20–40 per cent of the R&D value, while traditional farmers generate 60–80 per cent of the value (the private sector does not operate within this system). In addition, the table shows that land use under such an agricultural system is allocated stereotypically between non-intensive uses (80 per cent) and non-use or reserve (20 per cent). A change to high-yielding agriculture increases the R&D contribution of the private sector (to 10 per cent) and the public sector (to 60 per cent) while reducing the contribution of traditional farmers (to 30 per cent). The land use changes in such a change are very significant, with intensive land use increasing from 0 to 80 per cent of agricultural lands, and the amount of traditionally farmed lands reducing from 80 per cent to 15 per cent or less. The table therefore illustrates the basic differences between these systems, and how changes between them alter land uses and the contributions from various segments of agricultural society.

The following sections will assess the environmental and socioeconomic impacts of GURTs along the lines of changes in the structure of R&D and in the patterns of land use induced by this biotechnology. In effect, it will compare the way in which GURTs will effect movements between the systems outlined in Table 11.1.

In sum, the answer to the question about the impact of GURTs is relatively simple. We believe that GURTs may be viewed as a technological fix that allows the R&D system under F1 hybrids to be extended to crops other than maize and sorghum (the outbreeding plant varieties). The general implications of this shift, as demonstrated in Table 11.1, is that a transfer of the F1 hybrid system for all of agriculture has significant impacts on both the distribution of R&D and the allocation of land use.

THE DISTRIBUTIONAL IMPACTS OF GURTs

In the first instance there is no reason to believe that enhanced appropriability is harmful to the interests of farmers. This is because the *direct effect* of this technique will correspond only to the appropriation of the increased value of the innovations contained within the GURT seed. Farmers will continue to have the ability to purchase normally reproducing seed, only it will be without the innovations contained within the GURT seed. The availability of this seed will constrain the price at which GURT seed can be marketed, and it will mean that it will only be the added value of the innovation it contains that will cause it to be valued more highly. Therefore, at first sight, there is no reason to believe that farmers could be made worse off through the introduction of these technologies. These technologies merely add an option that did not exist previously.

Table 11.1 R&D structure and land use patterns for existing agricultural R&D systems

System and typical crop	Share of R&D in terms of (1) productive contribution to R&D (%) (2) apropriation of R&D rents (%)			Land use pattern		
	Private plant breeders	Public R&D through CG & NARCs rents through consumers	Trad'l breeders & farmers	Non-intensive agricultural use (%)	Intensive agricultural use (%)	Reserves (%)
Traditional agriculture (e.g. millet)		20/40	80/60	80		20
High-yielding agriculture (e.g. rice)	10/5	60/70	30/25	15	80	5
Fl hybrid agriculture (e.g. maize)	60/60	20/20	10/20	5	90	5
GURT agriculture		To be assessed			To be assessed	

Sources: CIMMYT (1999), Srinivasan (personal communication).

However, there are significant *indirect effects* that may also result from the introduction of these technologies. For example, in the future, farmers refusing to purchase GURTs may be denied not only the single restricted use innovation (attached to a general use plant variety), but also the use of an entire series of past innovations that have never diffused into general agriculture. It is this potential impact on the diffusion of agricultural innovations that is the most problematic characteristic of this new technology. In order to see this, it is important to contrast how agricultural innovation diffused in the past with how it is likely to diffuse in the future.

In the past, innovations in plant varieties have diffused into general use within agriculture over time even if released as protected varieties, because of the capacity to undertake breeding activities making use of them.[3] Sometimes this breeding activity was undertaken by private individuals but very often it occurred within public institutions. In fact publicly funded institutions (both at the national and international levels) have poured substantial funding and efforts into ensuring that recent innovations are diffused across the developing countries.[4] These public agricultural research institutions have done this by taking observed innovative characteristics and breeding them into locally used varieties.

Thus, a big difference between GURTs and the previous appropriation systems is that GURTs capture the value of the innovative characteristics by maintaining control over the plant variety in which they are embedded. These distinctions between the 'innovative characteristics' and the 'plant variety templates' (in which they are embedded) are categorized as the 'software' and the 'hardware' components of modern plant breeding activities. To a significant extent, in the past, even when the software has been developed by private breeders, it has been possible for much of it to diffuse quickly and inexpensively across the developing world by means of its incorporation within different hardware.

The movement toward GURTs may restrict the diffusion of recent innovations across developing countries. In effect the software would become hardware-specific, and it would be entirely up to the discretion and motivation of the innovator to diffuse its innovation to all of the various parts of the world on which it would confer benefits. In other words, the most significant (albeit indirect) impact of this change in technology would be the potential elimination of a currently diverse R&D sector (farmers, public sector, private sector) and its replacement by a fairly homogeneous and highly concentrated private sector. It is possible then that the rate and extent of diffusion of future innovations in agriculture would occur only under the exclusive control and direction of the originator of the initial innovation.

This is problematic for two reasons.[5] First, with sufficient time and a significant number of innovations, the other currently existing suppliers of plant varieties (private and public) might be rendered commercially obsolete. This would be the case if the alternative suppliers could only acquire the characteristics at high prices, and thus were only able to supply inferior substitute varieties. Then the farmer might face a small number of suppliers of viable seed, and consequentially much increased prices for GURT seed varieties.[6] Secondly, and if this were the case, the private sector might be able effectively to eliminate the public sector from all breeding activities on account of the need for licences and the restrictiveness of material exchanges. This might have deleterious consequences for those countries that are most highly dependent on public investment for their plant breeding needs.

Another perspective on this part of the problem is to note that the commercial sector may not have sufficient private incentives to diffuse their software widely; that is, across a diverse enough array of plant varieties.[7] Then the innovations would be aimed only at those markets where there was adequate demand, while general diffusion would be disallowed in order to protect those markets. The farmers on the fringes would be faced with farming with the innovative characteristics embedded in poorly performing varieties, or farming with the best local varieties but without the innovative characteristics.

Therefore the move to GURTs as an appropriation system is important primarily for the indirect effects that it might have on the entire system of R&D currently existing within agriculture. And these indirect effects are important because they might make a tremendous difference in the rate of diffusion of innovations to particular developing countries. In the next sections we first attempt to segregate the various categories of countries, and then we attempt to assess the extent to which the rate and direction of innovation will be slowed in those countries that are most affected.

THE ROLE OF DIFFUSION IN DETERMINING DISTRIBUTION

New technologies will have differing effects depending on the varying rates at which their benefits diffuse. The above discussion indicates that there are a few key factors that will determine the impact of these new technologies on the diffusion of benefits to various developing countries. The first factor is the capability of the developing country to undertake its own biotechnology. If it is able to do this, there is little change in the rate of technological diffusion with the introduction of GURTs, as these countries will be able to 'reverse engineer' GURT varieties as easily as they could any other variety.[8] For these countries (the biotechnology-capable) the impacts of GURTs are primarily positive. So the first important question for ascertaining the impact of this new technology is: *Does the subject country possess actual or incipient biotechnological capabilities?*

For countries without biotechnological capabilities, the important question concerns the impacts of GURTs on the rate of diffusion of innovations within their agriculture. In order to address this issue, we turn to an analogue from the agricultural industry. About 50 years ago the agricultural industry experienced its first technological revolution with the introduction of modern hybridized varieties (into sexually reproducing crops such as maize). To a large extent the advent of GURTs simply extends the effects of these forms of technologies to asexually reproducing crops (such as wheat and rice). The second important factor for ascertaining the impact of GURTs is: *Does the subject country have a significant investment in crops that are amenable to GURTs (such as wheat and rice)?*

The third important question will concern the extent to which the biotech-capable countries will include another country within their research strategies. *How quickly will the benefits from the technology diffuse to the non-biotechnology country?* Again, the experience with the hybridized modern varieties is instructive. We are able to examine how quickly and how extensively an individual country has benefited from innovations in maize breeding.

The change in biotechnology affects the group most severely. This consists of countries that are not likely to command biotechnologies and their commercial application for the foreseeable future. This means that they will certainly be adversely affected by the shift in the share of R&D rents from farmers and consumers to plant breeders. With no significant R&D contributions that could be protected with the use of GURTs, the R&D balance will move against these countries. This adverse impact is captured in a case study of the impact of hybridization-based technologies on the diffusion of yield gains. This study is reported in Chapter 8 of this volume. The chapter sets out the projected impacts of GURTs by reference to the distributional impacts that have been seen in the hybrid crops. In this first instance, these differences are notable from the relative gaps in yield evident in hybrid and non-hybrid crops. (See Table 11.2.[9])

The case study on hybrid varieties in Chapter 8 demonstrates that there is a reduced rate of diffusion of innovations from the frontier to developing countries with genetic use restrictions, resulting in a relatively wider productivity gap. Public sector investments in technology transfer were unable to overcome the reduced rate of diffusion resulting from the introduction of hybrid varieties.

In part this failure of technology transfer was a result of the crowding

Table 11.2 Average difference between crop yields of developing countries and developed countries, 1999

Crop	Difference in average yield (%)
Millet	3
Wheat	0
Potatoes	−18
Soybeans	−24
Rye	−26
Seed cotton	−29
Barley	−37
Rice, paddy	−39
Maize	−58
Sorghum	−69

Source: FAO Agriculture Database.

out of the public sector from the field where the private sector operated most. There are two possible explanations for this. One is that the cost of public sector operations was higher owing to the restrictive technologies applicable to hybrid crops. The other is that the cost of public sector operations was only *relatively* higher; that is, that the public sector chose to put its resources only into those crops where its efforts would have greatest impact. In either event, it is clearly the case that public sector activities did not counteract the restrictive nature of use restriction technologies, and the distributional implications for the less developed countries are clear.

In short, the availability of hybrid varieties in maize increased private plant breeding activities but reduced involvement in maize plant breeding by the public sector. This outcome indicates that the integration of the 'software' and 'hardware' of plant breeding makes the private sector more profitable, while rendering the involvement of the public sector less feasible. One explanation is that the public sector no longer offers a generally competitive alternative to the private sector's plant varieties, and so it leaves the field to the private sector.

The outcome is a widening gap between those countries whose plant breeding is on the technological frontier and those whose breeding lags behind. Clearly, the private sector is not putting the same amount of effort into diffusing its innovations throughout the developing countries as is the private/public mixed R&D sector for those crops unable to be hybridized. Just as clearly, the public sector is not able to continue to function in these markets, probably by reason of the restricted flow of materials and characteristics. The replacement of the mixed system of R&D by the private sector version has resulted in a systematic widening of productivity gaps between rich and poor countries.[10]

It may be possible to avoid the experience in the maize sector with the advent of GURTs. The advance of the private hybrid sector has taken place against the background of a policy of declining emphasis on public sector agricultural R&D. The growth in spending on agricultural R&D has been slowing down in developed countries, thus giving rise to a slowing down in the rate of diffusion. Global agricultural R&D spending disaggregated by developed and developing countries is presented in Table 11.3.

Table 11.3 Real agricultural R&D spending in developed and developing countries, 1971–91

	Expenditures (millions of 1985 international dollars)		
	1971	1981	1991
Developing countries (131)[a]	2984	5503	8009
Sub-Saharan Africa (44)[a]	699	927	968
China	457	939	1494
Asia and Pacific (excl. China) (28)[a]	861	1922	3502
Latin America and Caribbean (38)[a]	507	981	944
West Asia and North Africa (20)[a]	459	733	1100
Developed countries (22)[a]	4298	5713	6941
Total (153)[a]	7282	11217	14951
	Average annual growth rates (per cent)		
	1971–81	1981–91	1971–91
Developing countries	6.4	3.9	5.1
Sub-Saharan Africa	2.5	0.8	1.6
China	7.7	4.7	6.3
Asia and Pacific (excl. China)	8.7	6.2	7.3
Latin America and Caribbean	7.0	−0.5	2.7
West Asia and North Africa	4.3	4.1	4.8
Developed countries	2.7	1.7	2.3
Total	4.3	2.9	3.6

Note: [a] Figures in parentheses indicate the number of countries included in the respective totals.

Source: Alston *et al.* (1998).

The combination of increasingly complex technologies, increasingly restricted (and costly) material exchange and declining growth in public R&D investments makes it impossible for public sector plant breeders to maintain the breadth of objectives that existed previously. It is a combination of three factors – technological advance, material exchange restrictions and public investment levels – that will determine the overall impact on developing countries dependent on diffusion.

IMPLICATIONS FOR THE POOR WITHIN DEVELOPING COUNTRIES: FOOD SECURITY CONCERNS

Perhaps the most serious implication of GURT technologies lies in the degree of dependence introduced between the developing world and the biotechnology sector. For those countries without a well-developed biotechnology sector, there will be a much higher degree of insecurity arising out of their dependence on foreign firms for a continuing supply of seeds, and the R&D they contain. The other important implication of GURTs lies in their uneven distributional implications between sectors within developing countries, the same impact as they have between countries. To the extent that a developing country possesses both a modern and a traditional agricultural sector, the latter will fall even further behind the former with the introduction of GURTs.

Even if a developing country as a whole is set to benefit from the introduction of GURTs, an appraisal of the technology is incomplete without a closer examination of the distribution of its impacts within a country. There are strong a priori reasons for expecting that resource-poor farmers will be excluded from the benefits of technological progress in the direction of gene protection and that some will even be put at a disadvantage by this process.

Small-scale and subsistence farmers often form one of the largest groups in a developing country. Almost three-quarters of the poorest 20 per cent in Latin America, 57 per cent in Asia and 51 per cent in Africa can be found on low-potential lands (Leonard *et al.*, 1989). Their agriculture is on what is termed resource-poor and marginal lands with inadequate rainfall and adverse soil conditions. Fertility and topography limit productivity and chronic land degradation is widespread. Crops have to be adapted to particular local conditions of drought, poor soils and lack of irrigation.

There is evidence from the study of the experience of hybrid varieties that the significant private investment in maize improvement has consistently failed to reach resource-poor farmers (see Chapter 8 of this volume). There are two dimensions to this problem: development of agricultural productivity and distribution of income and wealth. These dimensions are discussed in the following sections.

Agricultural Productivity

Table 11.4 shows the impact of GURTs on the rate of productivity growth on farmers as a function of the speed of diffusion. The maize case study indicates that the diffusion of innovations in maize was slowed for the agricultural sectors furthest from the technological frontier. This will also be the case for GURTs, the maize case study suggests, if GURTs function by extending the impacts seen with hybrid crops. This means that the impact of GURTs (on individual farmers) will depend on the rate at which it causes innovation to diffuse to them, and on the GURT potential of the crops they grow. Since resource-poor farmers are also those most likely to lie far off the technological frontier (using traditional techniques and varieties), they are also those most likely to suffer negative or indifferent impacts from GURTs.

Table 11.4 Impact of diffusion rates on the rate of productivity growth for resource-poor farmers

Speed of diffusion of crop improvement in GURT crops	Low GURT potential	High GURT potential
Slow	Indifferent	Negative
Moderate	Indifferent	Indifferent
Rapid	Positive	Positive

This reveals that the group most affected by the negative impact of GURTs on the rate of productivity growth is the group of resource-poor farmers raising crops with high GURT potential. They will suffer in comparison with the current situation if the introduction of GURTs brings about a relative slowing of the rate of diffusion to them. The reduced rate of diffusion means that those furthest from the technological frontier receive the least amount of benefit from the innovation.

Distributional Impacts within a Developing Country

As the experience with hybrid crops suggests, technological protection of crop improvement creates strong distributional effects within developing countries through what has been termed the 'dualism' of systems. What is meant by this term is that some parts of the agricultural sector are comparatively advanced in terms of technology and crop mix, cultivate on relatively fertile soil and have access to credit, labour and capital markets. By contrast, remote areas with low-potential agricultural lands and little access to markets are characterized by a predominance of small-scale and subsistence agriculture.

It can be expected that the advanced parts of a developing country's agriculture will benefit from the introduction of GURTs to the same extent as they have from the introduction of hybrid crops (see Srinivasan and Thirtle, this volume). This will widen the distributional gap in developing countries to the extent that resource-poor farmers with a low GURT potential will not lose in absolute terms as shown above, but will lose relative to the developed sector. This implies that the dualism in these countries will be further exacerbated. In the maize case study this is indicated by the fact that the implicit productivity gap within developing countries is often larger than it is between countries. Table 11.5 shows that in Mexico, for example, the productivity gap in maize (difference between Mexico's yield and the yield at the technological frontier) is only 68 per cent, while it is 84 per cent between the modern and the traditional sectors within Mexico. The distributional implications of use restriction technologies are even more severe within developing countries than between developing and developed countries.

Table 11.5 Imputed within-zone yield gaps for Latin America, 1996

	Maize area	Improved hybrids (share of maize area, %)	Mean productivity gap developing to developed country (%)	Imputed within-country yield gap (between modern and traditional sector) (%)
Mexico	7.9	19.2	−68	−84
Central America	1.6	18.5	−78	−95
Andean zone	2.3	36.2	−69	−93
Southern cone	17.0	56.8	−45	−96

Source: Computed from CIMMYT (1999).

Food Security Impacts within a Developing Country

There are differential impacts of GURTs between different farmers in a country by virtue of the loss of fungibility of the harvest as both food and seed. The reason for this lies in the additional steps in the agricultural production process that are necessitated by GURTs. These steps require a presence and stability of infrastructure, markets and resource base that cannot be taken for granted for many resource-poor farmers.

Table 11.6 shows typical features of the advanced and resource-poor sectors in a developing country with respect to key requirements for a secure functioning of the cycle of agricultural production and exchange with GURTs. It demonstrates that, typically, resource-poor farmers are exposed to risks at every additional step in the cycle of production and exchange that GURTs introduce.

Table 11.6 Profile of advanced and resource-poor sectors in a developing country with respect to key requirements in the cycle of agricultural production and exchange

	Good access to market	High liquidity of market	Choice of suppliers	Access to credit	Quality of infrastructure
Advanced sector	Yes	Yes	High	Good	High
Resource-poor farmer	No	No	Poor	Poor	Low

Figure 11.1 shows a comparison of the cycles of production and exchange without and with GURTs. It highlights the importance of additional steps of seed sales, monetization of the harvest, purchase of seeds and their effective delivery in terms of volume and timing. Currently, 80 per cent of crops are estimated to be grown from saved seed and thus follow the left-hand cycle. A shift to GURTs would imply a number of additional steps to be carried out. As Table 11.6 shows, the requirements imposed for the faultless functioning of this extended cycle are often unsatisfactorily met in resource-poor areas, thus giving rise to risks of crop failure.

Conclusions Regarding Distributional and Food Security Impacts of GURTs

At the very best, the distributional and food security implications of these technologies can only be neutral. They are technologies that are devised to

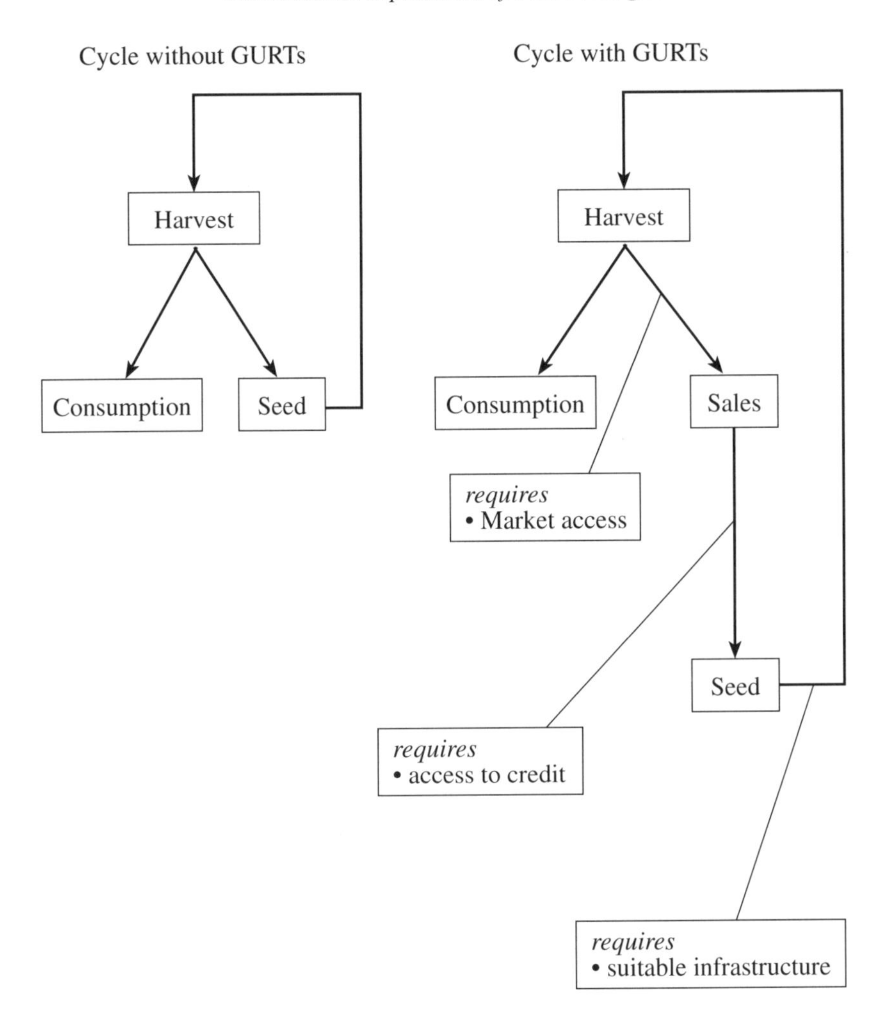

Figure 11.1 Cycles of production and exchange without and with GURTs

remove certain options from the typical farmer. For example, the farmer no longer has the option to self-supply seed in future years, or to undertake his own plant breeding activities. In exchange for this the farmer is meant to gain the option of increased rates of innovation produced by the biotechnology companies producing seed.

For the poor farmer in the poor country, the loss of options is almost certain to occur while the gain in rate of innovation is doubtful. This is because the biotechnology companies are unlikely to select the poor farmer in the poor country as the target for its innovations, or diffusion of innovations. Meanwhile, the poor farmer loses his options in order to protect the biotechnological investments that are focused elsewhere.

It is this fundamental asymmetry in direction of benefits and burdens that is most problematic in considering the distributional implications of these new technologies. The same caveats apply as before, however. If the poor developing country was to develop its own biotechnology sector, and focus it on providing innovations across its society, the poor farmer could gain as much as the modern one. The experience with F1 hybrids does not demonstrate that this has occurred; in fact, the private sector expansion has in some way crowded out those systems that were focused on the poor sectors of society (such as public sector breeders). It appears that the implications of these forms of technological change for the poor farmer are *relatively* unfavourable.

THE ROLE OF POLICY IN DETERMINING IMPACTS

The impact of these appropriation technologies will be dependent on the whole set of public policies that determine appropriation and diffusion. It is possible that the impacts of GURTs will differ significantly from those of the hybrids, if the public sector combines them with different funding and management policies.

Table 11.7 displays the rates of R&D diffusion for three different technology management systems and under two different assumptions concerning the level of public spending. For example, the table indicates that the rate of diffusion of innovations in hybrid varieties is slow with low levels of public funding (as the maize case study indicates) but is likely to be increased significantly if public sector funding were to be increased. On the other hand, the diffusion of innovations for non-hybrid varieties is

Table 11.7 Rate of R&D diffusion to countries dependent on funding (under different technology management systems and levels of public R&D spending)

Appropriation system		Public funding	
		Low	High
Current IPR	Hybrid varieties	Slow	Moderate
	Non-hybrid varieties	Moderate	Fast
IPR + GURTs		Slow	Moderate
GURT only		Slow/moderate	Fast

currently much more rapid and extensive than it is for hybrid varieties (but also dependent on public sector expenditures). Hence Table 11.7 indicates that agricultural R&D currently registers very different impacts on developing countries, depending on whether the variety is a hybrid one or not (cf. Srivastava *et al.*, 1996).

The critical issue for the impact of GURTs is whether (given the combination of management and spending policies applied) they will more closely follow the current example of the hybrid or the non-hybrid sector.

What are the different appropriation systems and what are their potential implications for diffusion? Table 11.8 explains the two different forms of appropriation thought possible for GURTs, in contrast to the current system of IPR protection that forms the baseline scenario. GURTs are

Table 11.8 Baseline and other scenarios for appropriation systems in agriculture

System	Description
Current IPR-based system (baseline)	This is the current system, where the release of GURTs is delayed indefinitely. It features one sector where legal protection through IPRs must be sought to protect R&D investment while it offers technological protection for a small set of outbreeding crops through hybridization. Under the current system, IPR remains the sole source of protection for R&D inputs into crop improvement in wheat, rice, cotton and other staple crops.
Combined use of GURTs and IPRs	This scenario regarding the legal implementation of GURTs can be considered as probable, as it does not require any changes in the existing legal structure. The situation would closely resemble the one that presents itself currently with combined legal protection in the form of *plant variety protection* (PVP) and technological protection through the use of *hybrid cultivars*. The introduction of GURTs would extend the feasibility of these combined regimes (beyond outbreeding crops). It is likely to have the effect of restricting the vast majority of agricultural R&D to the private sector, and of slowing the diffusion of innovation to developing countries in Group D.

Table 11.8 (continued)

System	Description
GURTs only – public sector policies directed to diffusion	It is possible for a combination of public sector policies to make the advent of GURTs a good system for both producing innovations and diffusing them. This would be the case if public policies were directed to the purpose of speeding the diffusion of private sector innovations (protected only by GURTs) throughout general agriculture, and GURTs were implemented in a manner that would not unnecessarily constrain diffusion. For example, this might be accomplished by means of mandatory licensing of GURT varieties to public sector institutions in regions where the private sector is not disseminating the varieties. Further research is required, but with adequate public sector spending it might be possible to encourage diffusion through focused policy interventions.

positioned to become an additional technological solution to the problem of rent appropriation. They represent the capacity to extend the solution concept of hybridization to those crop varieties that are not outbreeding. This suggests that the rates of diffusion for hybrids will be extended to those crops that are currently protected only through IPRs. This in turn implies a significant decline in the rate of diffusion of innovations across agriculture, which itself suggests a problem for those developing countries identified previously as dependent on the public sector for the diffusion of agricultural innovations.

The only real alternative is to meet this technological advance with public policies addressed to its deficiencies. These will be public policies that are directed to the purpose of aiding the diffusion of these innovations from the private sector to the public sector, and then on to those developing countries with little biotechnological capability.

NOTES

1. At present, GURTs are defined by patents and patent applications of four firms: USDA/Delta Pine & Land patent, US no. 5,723,765; Monsanto patent, no. WO 9744465; AstraZeneca patent, no. W09735983; Novartis patents for the creation of chemically-dependent plants with proprietary inducible promoters and genes.

2. The purpose of Table 11.1 is not to assert that these figures are concrete and actual estimates of the underlying parameters, but rather to illustrate how different R&D systems perform the allocation function differently. Then these figures will allow comparison with similar figures to be estimated for GURT-based R&D.
3 There are several important distinctions between PBR systems and GURT-based systems. First, PBR systems are limited in duration, while GURTs are not. Second, PBR systems often contain an 'own use' exemption for farmers that enable their own breeding activities. Third, and most importantly, PBR systems merely disallow the marketing of the same plant variety in competition with its innovator; they do not disallow breeding activities making use of the new plant variety (for example to translocate its innovative characteristics to other plant varieties).
4. The domestic system used for accomplishing this purpose is known as the national agricultural research centres (NARCs) while the international system devised in part to achieve the same purpose is known as the Consultative Group for International Agricultural Research (CG system).
5. Again, it is important to emphasize that there is nothing problematic in the first instance about the appropriation of the returns from innovation – this will simply enhance the prospects for effective investments in plant breeding activities. The problems that must be considered are second-order ones: the impact of wholly vesting agricultural R&D within the private sector and within a concentrated industry.
6. This would be the result of the refusal to licence innovations at reasonable prices to potential competitors, and the maintenance of very low prices until they were removed from the market. It is essentially the sort of conduct with which Microsoft has been charged.
7. For example, one well-known problem in the economics of market failure is the difficulty imposed by minimum efficient scales of operation. This can result in concentrating on the median consumer, or in this case the typical agricultural producer, and the failure to address the diversity of needs within a diverse consumer base.
8. The GURT form of system has little effect on biotechnology-capable countries, precisely because they will be able to use biotechnology to unravel and relocate the innovative characteristic. In this case the GURT merely provides a short-term advantage to the innovative breeder, possibly a head start of only two or three years in the marketing of the characteristic. The problem lies more in those countries with both little of their own biotechnological capacity and little investment by others interested in diffusing innovative characteristics into their local varieties.
9. The table shows the differences in the average yields of developing and developed countries weighed by area harvested.
10. It must be noted that this is both the most crucial conclusion of this report and also the most controversial. The implications for the poor people of developing countries of Table 11.2 of the extension of use restriction technologies are not favourable. Further research on this point is required to ascertain the nature of the relationship between Fl hybrids and technological innovation and diffusion.

REFERENCES

Alston, J.M. and R.J. Venner (1998), 'The Effects of U.S. Plant Variety Protection Act on Wheat Genetic Improvement, paper presented at the symposium on 'Intellectual Property Rights and Agricultural Research Impact' sponsored by NC 208 and CIMMYT Economics Program, 5–7 March, El Batan, Mexico.

Alston, J.M., P.G. Pardey and Vincent H. Smith (1998), 'Financing Agricultural R&D in Rich Countries: What's Happening and Why?', *The Australian Journal of Agricultural and Resource Economics*, 42(1), 51–82.

Butler, L.J. (1996), 'Plant Breeders' Rights in the U.S.: Update of a 1983 Study', in J. Van Wijk and Walter Jaffe (eds), *Proceedings of a seminar on 'The Impact of Plant Breeders' Rights in Developing Countries' held at Santa Fe Bogota, Colombia, March 7-8, 1995*, Amsterdam: University of Amsterdam, pp. 17–33.

Butler, L.J. and B.W. Marion (1985), *The Impact of Patent Protection on the U.S. Seed Industry and Public Plant Breeding*, Food Systems Research Group Monograph 16, University of Wisconsin Madison, Madison, WI.

Capalbo, S.M. and J.M. Antle (1989), 'Incorporating Social Costs in the Returns to Agricultural Research', *American Journal of Agricultural Economics*, 71, 458–63.

CIMMYT (1999), *A Sampling of Impacts 1999. New Global and Regional Studies*, Mexico: CIMMYT.

Conway, G. and G. Thoenissen (2000), 'Biotechnology, Food and Nutrition', *Nature*, Millennium Supplement.

Crouch, M. (1998), 'How Terminator Terminates', rev. edn, Edmonds Institute Occasional Papers, Edmonds Institute, Washington.

Evenson, R.E. (1989), 'Spillover Benefits of Agricultural Research: Evidence from US Experience', *American Journal of Agricultural Economics*, 71, 447–52.

Falck-Zepeda, J.B. and Greg Traxler (1998), 'Rent Creation and Distribution from Transgenic Cotton in the U.S', paper prepared for the symposium 'Intellectual Property Rights and Agricultural Research Impacts', sponsored by NC-208 and CIMMYT Economics Program, 5–7 March, CIMMYT Headquarters, El Batan, Mexico.

Frey, K.J. (1996), 'National Plant Breeding Study – I: Human and Financial Resource Devoted to Plant Breeding Research and Development in the United States in 1994', Special Report 98, Iowa Agriculture and Home Economics Experiment Station, Iowa State University, Iowa.

Fuglie, K., Nicole Ballenger, Kelly Day (1996), 'Agricultural Research and Development: Public and Private Investments Under Alternative Markets and Institutions', AER-735, Economic Research Service, United States Department of Agriculture.

Gupta, A. (1999), *Biosafety in an International Context*, Cambridge, MA: Harvard University Press.

Huffman, W.E. and R.E. Evenson (1993), *Science for Agriculture: A Long-Term Perspective*, Ames, IA: Iowa State University Press.

Jaffe, W. and J. Van Wijk (1995), 'The Impact of Plant Breeders' Rights in Developing Countries: Debate and Experience in Argentina, Chile, Colombia, Mexico and Uruguay', Inter-American Institute for Co-operation in Agriculture and University of Amsterdam, Amsterdam.

James, C. (1998), 'Global Review of Commercialized Transgenic Crops: 1998', ISAAA Brief no. 8. Ithaca, NY.

Jarvis, D. and T. Hodgkin (eds) (1998), *Strengthening the Scientific Basis of In Situ Conservation*, Rome: IPGRI/FAO.

Jefferson, R.A. with D. Byth, C. Correa, G. Otero and C. Qualset (1999), 'Genetic Use Restriction Technologies. Technical Assessment of the Set of New Technologies which Sterilize or Reduce the Agronomic Value of Second Generation Seed as Exemplified by U.S. Patent no. 5,723,765 and WO 94/03619', expert paper prepared for the Secretariat of the Convention for Biological Diversity, Subsidiary Body on Scientific, Technical and Technological Advice, Montreal, Canada.

Johnson, B. and K. Holmes (1998), *The 1999 Index of Economic Freedom*, New York: Dow Jones and Co.

Kalton, R.R. and P.A. Richardson (1983), 'Private Sector Plant Breeding Programmes: A Major Thrust in U.S. Agriculture', *Diversity*, 1(4), 16–18.

Kalton, R.R., P.A. Richardson and N.M. Frey (1989), 'Inputs in Private Sector Plant Breeding and Biotechnology Research Programs in the United States', *Diversity*, 5(4), 22–25.

Komen, J. and G.J. Persley (1993), 'Agricultural Biotechnology in Developing Countries: A Cross-Country Review', ISNAR Research Report no. 2. The Hague: International Service for National Agricultural Research.

Leisinger, K. (1995), 'Sociopolitical Effects of New Biotechnologies in Developing Countries', 2020 Vision Discussion Paper 2, IFRPI, Washington, DC.

Leisinger, K. (1997), 'Ethical and Ecological Aspects of Industrial Property Rights in the Context of Genetic Engineering', paper presented at Interlaken Conference.

Leonard, H.J with M. Yudelman, J.D. Stryker, J.O. Bowder, A.J. de Boer, T. Campbell and A. Jolly (1989), *Environment and the Poor: Development Strategies for a Common Agenda*, New Brunswick: Transaction Books.

Lesser, W. (1990), 'Sector Issue II: Seeds and Plants', in W.E. Siebeck (ed.), *Strengthening Intellectual Property Rights in Developing Countries: A Survey of Literature*, Madison, WI: University of Wisconsin Press, pp. 59–68.

Perrin, R.K., K.A. Kunnings and L.A. Ihnen (1983), 'Some Effects of the U.S. Plant Variety Protection Act of 1970', Economics Research Report no. 46, Department of Economics and Business, North Carolina State University.

Pray, C.E., M.K. Knudson and L. Masse (1993), 'Impact of Changing Intellectual Property Rights on US Plant Breeding R&D,' mimeo, Department of Agricultural Economics, Rutgers University, New Brunswick.

Pray, C.E., S. Ribeiro, Rolf A.E. Mueller *et al.* (1991), 'Private Research and Public Benefit – The Private Seed Industry for Sorghum and Pearl-Millet in India', *Research Policy*, 20(4), 315–24.

Qaim, M. and J. von Braun (1998), 'Crop Biotechnology in Developing Countries: A Conceptual Framework for Ex Ante Economic Analysis', ZEF Discussion Papers on Development Policy, no. 3. ZEF, Bonn.

Reilly, J. and T. Phipps (1988), 'Technology, Natural Resources and Commodity Trade', in John D. Sutton (ed.), *Agricultural Trade and Natural Resources: Discovering Critical Linkages*, Boulder, CO: Lynne Rienner Publishers, pp. 224–35.

Rejesus, R., M. van Ginkel and M. Smale (1998), 'Wheat Breeders' Perspectives on Genetic Diversity and Germplasm Use. Findings from an International Survey', WPSR no. 40, CIMMYT.

Schmidt, J.W. (1984), 'Genetic Contributions to Yield Gain in Wheat', in W.R. Fehr (ed.), *Genetic Contributions to Yield Gains in Five Major Crop Plants*, Madison: Crop Science Society of America, pp. 89–101.

Smale, M. (1997), 'The Green Revolution and Wheat Genetic Diversity: Some Unfounded Assumptions', *World Development*, 25, 1257–69.

Southgate, D. (1997), 'Alternatives to the Regulatory Approach to Biodiverse Habitat Conservation', *Environment and Development Economics*, 2(1), 106–10.

Srivastava, J., N. Smith and D. Forno (1996), 'Biodiversity and Agriculture. Implications for Conservation and Development', World Bank Technical Paper no. 321, World Bank, Washington, DC.

Swanson, T. (ed.) (1995), *The Economics and Ecology of Biodiversity Decline. The Forces Driving Global Change*, Cambridge: Cambridge University Press.

Thirtle, C.G. (1985), 'Technological Change and Productivity Slowdown in Field Crops: United States, 1939–1978', *Southern Journal of Agricultural Economics*, 17, 33–42.

Thirtle, C.G., V.E. Ball, J.C. Bureau and R. Townsend (1994), 'Accounting for Efficiency Differences in European Agriculture: Cointegration, Multilateral Productivity Indices and R&D Spillovers', mimeo, University of Reading.

Traxler, G., J. Falck-Zapeda, J.I. Ortiz-Monasterio, and K. Sayre (1995), 'Production Risk and the Evolution of Varietal Technology', *American Journal of Agricultural Economics*, 77, 1–7.

Tzotzos, G. (1995), 'Genetically Modified Organisms: A guide to Biosafety', Wallingford: CABI.

Virchow, D. (1999), *Conservation of Genetic Resources: Costs and Implications for a Sustainable Utilization of Plant Genetic Resources for Food and Agriculture*, Berlin: Springer Verlag.

White, F.C. and J. Havlicek (1979), 'Rates of Return to Agricultural Research and Extension in the Southern Region', *Southern Journal of Agricultural Economics*, 11, 107–11.

World Conservation Monitoring Centre and Faculty of Economics (1996), 'A Survey of the Plant Breeding Industry, in T. Swanson and R. Luxmoore (eds), *Industrial Reliance upon Biodiversity*, Cambridge: WCMC.

Index

access, enabling technologies 101–4
accountability 126
AFLPs *see* amplified fragment length polymorphisms
Africa
  agriculture research 30–1
  see also sub-Saharan Africa
African Association for Biological Nitrogen Fixation 84
African Biosciences Network 84
African Intellectual Property Organization 92
African Regional Intellectual Property Organization 92
Agracetus 76
Agrarian Research Group 124
AgrEvo 76, 163
agribusiness 74
agricultural biotechnology
  advances 3–10
  capital concentration 74–8
  conclusion 124–6
  distribution of benefits 67–72
  incentives for diffusion of benefits 109–24
  intellectual property rights 88–100
  investment 72–4
  orphan crops 84–6
  proprietary rights 100–9
  research collaboration 78–84, 125
  research institutions, Africa 232
  social venture capital 86–8
agricultural extension service 238
agricultural intensification, and population
  biotechnology 35–41
  challenges and successes 26–9
  failures 29–33
  future trends and challenges 33–5
agricultural R&D 45–7, 250
  changes in 118–19
  current mixed system 56–9
  impact of GURTs 59, 250–2
    determining net impact 62–3
    developing countries as a group 61–2
    individual developing countries 63–4
    on system 60–1
  intellectual property rights
    distribution of rents 52–3
    incentive to supply 53–6
  land use patterns 251–2, 253
  plant breeding 11–12
  private sector, OECD countries 73
  public sector component 44
  segregating between contributions 49–50
  spending, developed/developing countries 258
agricultural research centres, international 163
agricultural systems, stereotypical 251–2
Agroceres 76
agrochemical companies 75
amplified fragment length polymorphisms 221, 222
anti-biotechnology lobbying 75, 126
antibiotics 165
antitrust enforcement policies 77
appropriability
  impact of GURTs 60
  plant breeding industry 7–8
  private sector plant breeding 154
  seed companies 170–6
  technological solution 8–9
  terminator technology 156–9
appropriation systems, implication for diffusion 265–6
Argentina, plant breeders, R&D expenditure 55
Asgrow v Winterboer 157

Asia
  food production 28–9
  green revolution 32
asset ownership
  green revolution 33
  successful irrigation 32–3
ASSINSEL *see* International Association of Plant Breeders for the Protection of Plant Varieties
ASTRA Zeneca 76, 78, 163
Aventis Agriculture 76

Bayh-Dole Act (1980) 80
biomedical research 85
biotechnology
  as an industrial movement 3–10
  capability, developing countries 180–1
  impacts of 10–13
  population, and agriculture intensification 35–41
  sub-Saharan Africa 230–43
  see also agricultural biotechnology; molecular genetics
Boserup model 26–7, 30
brain drain, developing countries 121
brand-name drugs 114
breeders' exemption 96
Breeders Ordinance (1941) 155
breeding *see* plant breeding; pre-breeding; traditional breeding
'brown bag' sales 157

Calgene 76, 95
CAMBIA *see* Centre for the Application of Molecular Biology to International Agriculture
Canada, patents
  compulsory licences 108
  research exemptions 96
capacity building, molecular genetics 226
capital concentration, agricultural biotechnology 74–8
Carlsberg and Danisco 75
Cassava Biotechnology Network 84, 110, 112
Centre for the Application of Molecular Biology to International Agriculture 112, 113, 118
Centre de Coopération Internationale en Recherche Agronomique 80
Centre National de la Recherche Scientifique 80
CGIAR *see* Consultative Group on International Agriculture Research
Chakraborty v Diamond 53, 158
child labourers 70
Chilean Agricultural Research Institute 123–4
chromosome 22, 220
client-driven plant biotechnology research 122–4
clonal propagation, Africa 232
Codex Alimentarius 84
Cohen-Boyer patent 104
collaboration, agricultural biotechnology 78–84, 125
Commission on Genetic Resources (FAO) 84
commoditization, of seed 154
competitive pressure, innovations 39
compulsory licensing, patents 107–9
Congressional Budget Office (US) 115
conservation
  biotechnology 219
  molecular genetics 221–3, 224
Consultative Group on International Agriculture Research 12, 56, 59, 73, 83, 90, 104, 113, 118, 164
contracts, agricultural biotechnology 94–5
Convention on Biological Diversity 82, 84, 95
copylefting 111–12
copyright 6, 142
cre-loxP recombination technology 103
crops
  yields, developing/developed countries 257
  see also hybrid varieties; non-hybrid crops study
cross-licensing, patents 98
cultivars, genetic transformation 225–6

databases, IPR changes 103
deadweight losses, IPR, new plant varieties 139, 140
defensive patenting 112
DeKalb 76
demand, productivity 33–4
developed countries
  agricultural R&D, public sector 56–7
  PVP legislation 156
developing countries
  agricultural biotechnology 67–126
  agricultural R&D
    public sector 56–7
    spending 58, 258
  crop yields 257
  impacts of GURTs
    as a group 61–2
    implications for poor 259–64
    individually 63–4
    non-hybrid varieties 206–8
    preliminary assessment 177–94
  PBR titles 55
  population and agricultural intensification 25–41
  technological change 19–20
  technology transfer 118–21
  terminators
    framework for economic analysis 150–67
    legal analysis 137–47
diffusion
  appropriation systems 265–6
  hybrid variety study 201–5
  incentives for 109–24
  role in distribution 255–9
direct effects, GURTs 178
disclosure rules, terminator technology 147
disease resistance, terminator technology 165
disease-free plant material 232
distinctness, plant varietal protection 93, 155
distribution
  impact of GURTs 252–5, 260–2
  role of diffusion in 255–9
  technological change 13–19
diversity, molecular studies 224
DNA banks 225
DNA forensics 97
DNA Plant Technology 76
DNA sequences
  data on 222
  molecular markers 221
  patents 103
dormant traits 151
Dow Agrosciences 76, 163
Dow Elanco 76
downstream seed companies 75
drugs, generic 114–15
DuPont 75, 76, 103, 163
durable monopoly problem, plant breeders 4–5
'durable goods monopolist' 3

ecology
  adoption of new varieties 240–1
  terminator varieties 164–5
economic power, IPR regimes 140
efficiency, agricultural R&D 59
employment, privatization of research 81
Empresa de la Moderna 74, 76
enabling technologies, patents and access 101–4
enforcement problem, plant breeders 7–8
essential derivation 96, 97, 156–7
ESTs *see* expressed sequence tags
Eurasian Patent Convention 92
Europe, plant varietal protection 138
European Patent Convention 91, 92
European Science Foundation 104
European Union, funding, agricultural biotechnology research 81
expressed sequence tags
  gene expression 222
  patents 103
extension officers 238
externalities, biotechnology 37

F1 hybrid-based agriculture 251, 252
FAO *see* Food and Agricultural Organisation
farm-saved seed
  right to save and regrow 99–100
  sale of 157
  sterility of non-terminator varieties 164

farmers
information, natural selection 48
participation in research agendas 126
see also resource-poor farmers
farmers' privilege 99–100, 156
farmers' rights 6, 11–12, 53
Federal Trade Commission (US) 77
'first to file' basis 91
'first to invent' basis 91
FIS *see* International Seed Trade Federation
FlavrSavr tomato 93
Food and Agricultural Organisation 6, 33, 84
food availability, population growth 26
Food and Drug Administration 115
food production, Asia 28–9
food security
agricultural biotechnology regulation 78
farmers' privilege 100
impacts of GURTs 262–4
research exemptions 125
foreign direct investment, terminator technology 166
Free Software Foundation 111
freedom to operate 163, 165
funding
agricultural biotechnology 80–1
push-pull strategy 86–8
plant biotechnology 84
research bias 74

Gatsby Charitable Foundation 118
GATT (General Agreement on Tariffs and Trade) 6, 90
GAVI *see* Global Alliance for Vaccines and Immunisation
gender bias, adoption, new varieties 239
genebanks
management 224–5
molecular genetics 222–3
generics industry 113–17
genetic maps 221–2
generic material transfer agreements 98–9
genetic resources, R&D 46
genetic use restriction technologies *see* GURTs
genetically modified (GM) crops 220
IPR 137
side effects 37
genome sequencing 221
Genoplante 80
germplasm
pre-breeding 222
security officers 97
Ghana, maize project 237–8
Global Alliance for Vaccines and Immunisation 87–8
global population growth 25
GNU/Linux systems 111
Golden rice 78
green revolution
asset ownership 33
avoidance of mass starvation 230–1
innovations, demand and supply 28
intensive systems 34–5
seed replacement rate 165–6
success, Asia 32
'growth clubs' 211
GURTs
defined 8
impacts *see* impacts of GURTs
implications for poor 259–64
see also terminators

Hatch-Waxman Act (1984) 115–16
Hayami-Ruttan model 11, 27–8, 30, 31
herbicide resistant plants 35, 36
high-yielding agriculture 251, 252
Hoescht AG 76, 77
human capital investment, biotechnology 40–1
Human Genome Project 220
hybrid varieties
appropriating returns 56
growth in yields 160
investment in 55, 161
maize
adoption by poorer farmers 40
availability of 257
private sector plant breeding 154
hybrid variety study
panel study on diffusion 201–5
preview of results from 189–90

hybridization
  implications 200
  use restriction 7–8

IDRC Canada 232
ILTAB 111
impacts of GURTs 249–52
  agricultural R&D 59–64
  developing countries
    categories of countries 180–1
    classification 182–5
    conclusion 193–4
    diffusion problem 185–9
    factors determining 178–80
    hybrid variety study 189–90
    impact groups 181–2
    policy choices 190–3
    public sector spending 190
  distributional 252–5
    technical change 13–19
  non-hybrid crop varieties 205–12
incentive systems, intellectual property rights 51–2
indirect effects, GURTs 178, 254
induced innovation model 27–8, 30, 40
information
  agricultural R&D 53
  biotechnology, Africa 231–4
  from genetic resources 46
  from natural selection 48–9
  molecular genetics 226
INIA *see* Chilean Agricultural Research Institute
INIAA *see* Instituto Nacional de Investigacion Agraria y Agroindustrial
innovation possibility curve 28, 40
innovations
  agricultural biotechnology 100–1
  agriculture 3–4, 28
  competitive pressure 39
  traditional approach to generation of 47–9
input traits 71
Institut de Recherche pour le Développement 80
Institut National de la Recherche Agronomique 80
Institute for Functional Genomics 74
institutional adjustment, population growth 31–2
Instituto Nacional de Investigacion Agraria y Agroindustrial 123
Instituto Nacional de Tecnologia Agricola 123
Instituto Rio Grandese do Arroz 123
INTA *see* Instituto Nacional de Tecnologia Agricola
intellectual property rights
  Africa 240
  agricultural biotechnology 88–100
    claims relating to 38–9
    innovation system 100–1
  agricultural R&D
    distribution of rents 52–3
    incentives to supply 53–6
  genetically modified plants 137
  incorporated into seeds 35
  new plant varieties 137–9
    merits and demerits 139–41
  plant breeding 152–6
  public sector involvement 82
  rate of return on 49
  rewards, informational investment 51–2
  terminator technology 143, 162
  see also patents
intensive systems, green revolution 34–5
Intermediary Biotechnology Service 120
international agricultural research centres 163
International AIDS Vaccine Initiative 88
International Association of Plant Breeders for the Protection of Plant Varieties 97
International Association for Plant Tissue Culture 84
International Convention for the Protection of New Varieties of Plants 155
International Federation of Agricultural Producers (IFP) 121
International Patent Cooperation Treaty 92

International Plant Genetic Resources Institute 219, 223–4
International Rice Research Institute 50
International Seed Trade Federation 83, 99
International Service for the Acquisition of Agri-Biotech Applications 107
International Society for Plant Molecular Biology 84, 113
International Society for Tropical Root Crops 84
International Undertaking on Plant Genetic Resources 6, 53
Internet, copyright 142
invention, defined 91
investment
  agricultural biotechnology 72–4, 121–4
  agricultural R&D 53–6, 60
  compensation for 5
  human capital, biotechnology 40–1
  plant breeding 153–4
  terminator technology 166
IPGRI *see* International Plant Genetic Resources Institute
IPR *see* intellectual property rights
IR8 rice 236
IRGA *see* Instituto Rio Grandese do Arroz
irrigation, land ownership 32–3
ISAAA *see* International Service for the Acquisition of Agri-Biotech Applications
ISNAR 113
ISNAR-IBS 120
ISPMB *see* International Society for Plant Molecular Biology
IUPGR *see* International Undertaking on Plant Genetic Resources

John Innes Centre (UK) 80
joint ventures 77

Kenya, biotechnology research 232

label notices 157
labour/land ratio, population growth 26–7
Land Grant universities (US) 104, 113
land ownership 32–3
land use patterns, agricultural R&D systems 251–2, 253
Latin America, investment in R&D 54–5
Law on the Protection of Varieties and the Seeds of Cultivated Plants (1953) 155
lawyers
  patent 97, 101
  technology transfer 82
legal disputes 76–7
legal protection, new plant varieties 137–9
legislation, PVP 154–6
licensing
  patented biotechnologies 104–5
  of patents 81–2
life sciences companies 74, 163

maize
  genetic research, Africa 240
  market structure, adoption of new varieties 239
  productivity gap, Latin America 261
  see also hybrid varieties
maize project, Ghana 237–8
Malthus 26
market, agricultural biotechnology 70
market segmentation, patents 105–7
market structure, adoption of new varieties 239
marketing, agricultural biotechnology 71
material transfer agreements 98–9, 104, 110–11, 113
men, adoption, new varieties 239
mergers and acquisitions, agricultural biotechnology 74–8, 163
micropropagation companies 77
molecular genetics
  key issues and IPGRI 223–7
  opportunities for conservation and use 221–3
  power of 220–1
molecular markers 37, 221

molecular profiling, varieties 97
monopoly power, IPR 38–9
Monsanto 38, 74, 76, 77, 95, 98, 107, 157, 163
multinationals
  biotechnology sector 38, 74
  hybrid varieties 55
  life sciences companies 163
Mycogen 76

NAFTA 107
Napster 6
NARS *see* national agricultural research systems
National Academy of Sciences (US) 103
national agricultural research systems
  Africa 30–1, 240
  defensive patenting 112
  funding and research bias 74
  plant breeding 56
natural selection, plant varieties 48
needs, resource-poor farmers 69–70
net impact, GURTs 62–3
new plant varieties
  Africa
    adoption 237–41
    production 235–6
  durable monopoly problem 3–5
  legal protection 137–9
    merits and demerits 139–41
  natural selection 48
  registration 5, 53
  terminator genes 142
  uniform criteria 93, 155
Nigeria, adoption new soybean varieties 238
non-hybrid crops, growth in yield 160
non-hybrid crops study 205–12
  baseline scenario and parameters 205–6
  implications 211–12
  qualifications 212
  simulation results
    developing countries 206–8
    selected countries 209–11
non-terminator varieties, sterility 164
not-for-profit companies 88
Novartis 76, 80, 163
novelty, plant varietal protection 93

OECD countries
  agricultural R&D
    private sector 73
    spending 57
  anti-biotechnology lobbying 126
  patents 90
    lawyers 101
    portfolios 81
  plant biotechnology research 83
OECD Development Centre study 120–1
Offices of Research and Technology Applications 80
open pollinate varieties 238–9
orphan crops 84–6
orphan drug acts 85–6
output traits 71
'own use' provision 6, 59

Panel on Proprietary Science and Technology (CGIAR) 113
Paris Convention for the Protection of Industrial Property 92, 108, 153
partial price discrimination 144, 145
partnerships, agricultural biotechnology
  public-private 87–8
  strategic 76
Patent Cooperation Treaty 92
Patent Statute (US) 137
Patent and Trade Mark Office (US) 103
patentees, profit maximising behaviour 139, 140
'patent misuse' doctrine 144
patents 91–2
  compulsory licensing 107–9
  key enabling technologies 101–4
  licensing of 81–2
  OECD countries 90
  pharmaceutical industry 115
  plant biotechnology 93
  preferential access
    costs of licensing 104–5
    importance of segmented markets 105–7
  research exemptions 96–9
pathology, molecular 226
peasant farming, low input-low yield equilibrium 39–40

perfect price discrimination 161
pharmaceuticals, generic 114–15
Pharmacia & Upjohn Inc 76
Pharmacia Corporation 76
Pioneer HiBred 74, 75, 76, 93, 97, 157
piracy, marketing rights 51
plant biotechnology
  commercial potential 71
  patents 93
  products/techniques 100–1
  research
    OECD countries 83
    orphan crops 84–5
    publicly funded 79–80
  resource-poor farmers 110–13
Plant Biotechnology Network 84
plant breeders
  durable monopoly problem 4–5
  enforcement problem 7–8
  R&D expenditure, Argentina 55
plant breeders' rights 5–7, 11–12, 92–4
plant breeding
  agricultural R&D 11
  biotechnology 37
  investment 54–6
  IPR 152–6
  vertical structure 47–9
Plant Genetic Systems 76
Plant Patent Act (1930) 154–5
Plant Protection Act (US) 137
plant varietal protection 92–4
  farmers' privilege 99–100
  legislation 154–6
  research exemptions 96
plant varieties
  herbicide resistant 35, 36
  patents 91
  protection 52–3
  see also hybrid varieties; new plant varieties; non-hybrid crops
Plant Variety Protection Act (US) 137
policies
  impact of GURTs 190–3, 264–6
  molecular genetics 226–7
  public sector research institutions 79–80
policy instruments, agricultural biotechnology 125
poor, implications of GURTs 259–64
poor farmers *see* resource-poor farmers
population, and agriculture intensification
  biotechnology 35–41
  challenges and successes 25–9
  conclusion 41
  failures 29–33
  future trends and challenges 33–5
poverty
  adjustment to population growth 30, 39–40
  factors underlying rural 69
pre-breeding 222
preferential access
  importance of segmented markets 105–7
  proprietary agricultural biotechnologies 104–5
prescription drugs 114, 116
price controls, terminator technology 147
price discrimination, terminator technology 144–5
prior publication approach 110
private sector
  agricultural biotechnology
    collaboration with public sector 78–84
    investment 72–4
  diffusion of innovations 257–8
  investment
    agricultural R&D 53–4, 60
    plant breeding 154
pro-poor bias, agricultural biotechnology 68–70
PROARROZ 123
process needs, resource-poor farmers 70
product needs, resource-poor farmers 69–70
production function (R&D) 50
productivity
  demand growth 33–4
  implications of GURTs 260
  resource-poor farmers 69
profits, new plant varieties 143
property rights regime *see* intellectual property rights

proprietary rights, agricultural biotechnology 90, 100–9
proprietary traits 71
Protection of Biotechnological Inventions (EU Directive) 100, 103, 108
public domain plant biotechnologies 110–13
public sector
agricultural biotechnology
collaboration with private sector 78–84, 125
demand-driven research agendas 71
investment 72–4
agricultural R&D 44, 56–9
biotechnology research 39
crowding out of 256–7
impact of biotechnology 12
investment
diffusion of biotechnology's benefits 117–18
plant breeding 153–4
spending, GURTs 190
public support, agricultural R&D 119
pull funding strategies 87
purchase contracts 157
push funding strategies 87
PVP *see* plant varietal protection

quantitative trait loci 221

R&D *see* research and development
random amplified polymorphic DNA 221
rate of return
IPR 49
new products 51
reach-through licence agreements 103
reach-through royalties 102–3
regional trade agreements 107
registration, plant varieties 5, 53
regulation
agricultural biotechnology 77–8
terminator technology 147, 165
rent dissipation 140
rent distribution, IPR protection 52–3
reproduction technologies, copyright 6
research, molecular genetics 224
research and development
agricultural biotechnology
funding strategies 86–8
targeting local investments 121–4
rate of return, new products 51
see also agricultural R&D
research exemptions, patents 96–9
research institutions, Africa 232
research synergies 76
research tools, patents 101–4
research-management extension (RME) model 118
resource-poor farmers
agricultural biotechnology 67–8
economic dependency, terminators 145–6
farmers' privilege 100
participation, agricultural R&D 121–4
pro-poor bias, biotechnologies 68–70
public domain plant biotechnologies 110–13
restriction fragment length polymorphisms 221, 222
reward systems, agricultural biotechnology research 81
RFLPs *see* restriction fragment length polymorphisms
rice
genetic resources input, new varieties 50
intensive monoculture 35
IR8 variety 236
transgenic "Golden" rice 78
Rockefeller Foundation 73, 111
Roundup resistant plants 35, 36
Ruttan *see* Hayami-Ruttan

scientists
African, shortage of 233, 234
brain drain, developing countries 121
second-generation terminator technologies 142
security officers, germplasm 97
seed companies
appropriability, new varieties 170–6
downstream 75
investment in research 161

seed companies (*cont.*)
IPR litigation 158–9
M&A, developing countries 77
seed industry
improving appropriability 157
mergers and acquisitions 75–86, 163
seed replacement rate 165–6
seeds
biotechnology 38
commoditization of 154
sterility *see* terminator technology
world market for, crops 72–3
self-pollinating varieties
investment in 55
terminator technology 161
small companies, agricultural biotechnology 77
smallholders
adjustment to population growth 39–40
farmers' privilege 100
social venture capital, agricultural biotechnology 86–8
Sony v Amstrad 6
soybean, adoption new varieties, Africa 238
Special Programme on Biotechnology and Development Cooperation 118
spending, public sector R&D 57–8
stability, plant varietal protection 93, 155
State Science and Technology Commission (China) 123
sterility, non-terminator varieties 164
Stevenson-Wydler Technology Innovation Act 80
strategic partnerships, agricultural R&D 76
sub-Saharan Africa
biotechnology 230–43
activities by country 245–6
summary and conclusions 241–3
traditional breeding 234–41
Syngenta 76
systematic botany 154
Systemwide Programme on Participatory Research and Gender Analysis (CGIAR) 118
technological change
in agriculture 32
developing countries 19–20
distribution 13–19
technological solution, appropriability 7, 8–9
technological yield gap 34–5
technology premium 94
technology protection systems *see* GURTs
technology transfer
developing countries 118–21
lawyers 82
units 101
technology use fees 94
terminators 95
defined 8
developing countries
advantages 143–5
analysis of economic rationale 170–6
appropriability 156–9
choices for 164–6
concluding observations 166–7
disadvantages 145–6
economic implications 159–64
implications of technology 151–2
plant breeding and IPRs 152–6
potential functions 141–3
regulating usage 146–7
Third External Review Panel (CGIAR) 73
Third System Review (CGIAR) 83, 90
Third World Academy of Sciences 84
time-labour allocation, resource-poor farmers 70
tissue culture technologies 219, 232–3
Trade Related Intellectual Property Rights *see* TRIPS agreement
trade secrets 93
trademarks 93
traditional agricultural systems 251–2
traditional breeding, Africa 234–41
traditional farmers, information, natural selection 48
trait-based collections 222–3
trait-based GURTs 8
traits
identification of 221
searching for useful 225

Transatlantic Economic Partnership 90
transgenic approaches 37
transgenic crops
  concerns about 220
  global market for 72–3
  publicly funded research 78
TRIPS agreement 6, 56, 90, 92, 106–7, 108, 138, 164

uniformity, plant varietal protection 93, 155
Union pour la Protection des Obtentions Végétales 5, 6, 52, 59, 92–3, 96, 138, 155, 156
United States
  expenditure, agricultural R&D 54
  funding, agricultural biotechnology 80
  IPR protection 52–3
  patents
    'first to invent' basis 91
    research exemptions 96
  protection, new plant varieties 137
UPOV *see* Union pour la Protection des Obtentions Végétales
Uruguay Round (GATT) 6, 90
use restriction, hybridization 7–8
utility patents 97, 157

Vaccine Development and Purchase Funds 88
vaccines, research 87–8
variety-based GURTs *see* terminators
vertical industry 47–9

West Africa Rice Development Association 235–6
WIPO *see* World Intellectual Property Organization
women
  adoption of new varieties 239
  resource-poor farmers 69
Working Group on Research Tools 104
World Bank 73, 88, 119
World Health Organisation 107
World Intellectual Property Organization 92
World Trade Organisation 6, 84, 90, 109

Zeneca 76, 80